The Method of Functionals in the Quantum Theory of Fields

RUSSIAN TRACTS ON ADVANCED MATHEMATICS AND PHYSICS

Volume I

A. V. POGORELOV, Topics in the Theory of Surfaces in Elliptic Space

Volume II

A. I. KHINCHIN, Analytical Foundations of Physical Statistics

Volume III

G. F. KHILMI, Qualitative Methods in the Many Body Problem

Volume IV

M. A. EVGRAFOV, Asymptotic Estimates and Entire Functions

Volume V

YU. V. NOVOZHILOV and A. V. TULUB, The Method of Functionals in the Quantum Theory of Fields

Additional volumes in preparation.

The Method of Functionals in the Quantum Theory of Fields

by Yu. V. Novozhilov
and A. V. Tulub

With an introductory article

Equations with Variational Derivatives in Problems of Statistical Physics and of Quantum Field Theory

by N. N. Bogolyubov

GORDON AND BREACH
Science Publishers, Inc., New York

Originally published as

Metod functsionalov v kvantovoi teorii polya
Uspekhi Fizicheskikh Nauk LXI, 1, 53-102 (1957)

Uravneniya s variatsionnymi proizvodnumi v problemakh statisticheskoi fiziki i kvantivoi teorii polya
Vestnik Moskovskogo Universiteta 10, 115-124 (1955)

Library of Congress Catalog Card Number 61-18833

GORDON AND BREACH, Science Publishers, Inc.
150 Fifth Avenue, New York 11, New York

PRINTED IN THE UNITED STATES OF AMERICA

Preface

In recent years the method of functionals has been widely used in the quantum theory of fields. Much attention has been devoted to these methods in the hope of obtaining, by their use, exact solutions of some problems, or at least to find some exact relations in the theory of fields. Until recently practically all work in the quantum theory of fields has been done on the basis of the perturbation theory.

At present the majority of physicists share the conviction that even in quantum electrodynamics, where the coupling constant is small, the perturbation theory cannot serve as a basis for the study of matters of principle, because the approximation series may diverge. In meson theory, methods of solving the equations of the quantum theory of fields are urgently necessary, because the coupling constant of the nucleons and the meson fields cannot be considered as small.

The method of functionals, unlike other methods, allow us to formulate rigorously the equations for the field functions and make it possible to obtain a formal solution of the problem of interacting fields. This peculiarity of the method of functionals is important for investigations of a fundamental character as well as for the elaboration of approximation methods of solution of the field equations which are distinct from perturbation theory.

Work based on the method of functionals may be divided at present (from the point of view of using the calculus of functional) into two categories: the study of generating functionals and the study of the use of functional integration.

The idea of the method of the generating functional was put forward by V. A. Fock in 1928 [1], and its details were elaborated in his work in 1934 [2]. This method has been used in the following years for the solution of a number of problems [5-8], but it has been developed and used widely only in recent years as a result of the general development of the quantum theory of fields which gave first priority to the requirements of relativistic covariance of the field equations and to the possibility of solving them without the use of perturbation theory. The most important development of the method of functionals is related to the introduction by Schwinger [9] of the functionals of external sources which are the generating functionals for relativistic field functions and the equivalent Lagrangian formulation of Feynman [10].

Several papers on the method of functionals have been reviewed by Berestetskii and Galanin [3] and Silin and Feinberg [4]; therefore we shall devote relatively little space to the problems discussed in their articles.

In Part I of our survey we shall consider in detail the foundations of the method of functionals in the quantum theory of fields and the generating functionals for non-relativistic functions.

In Part II we shall consider problems relating to generating functionals for relativistic functions, functionals in space-time approach, and the integration of functionals over a Fermi field.

Contents

The Method of Functionals
in the Quantum Theory of Fields

EQUATIONS WITH VARIATIONAL DERIVATIVES IN PROBLEMS OF STATISTICAL PHYSICS AND OF QUANTUM FIELD THEORY

N. N. BOGOLYUBOV

(Delivered at the Lomonosov lectures on 14 April 1954)

The method of equations in variational derivatives, a method closely associated with methods of the type of secondary quantization, is studied in the article. A new operator interpretation of the Schwinger theory of Green's functions is given on the basis of this method. At the same time, a new "single-particle approximation", based on the neglecting of isolated loops in the Feynman diagrams, is formulated.

Equations are obtained similar in their form to the corresponding equations of ordinary single-particle theory, but substantially differing in commutation relations. Methods of the Bloch-Nordsieck, strong coupling, etc. type were used for their approximate solution.

The effectiveness of the method of equations in variational derivatives is also demonstrated for problems of the theory of the statistical equilibrium of classical systems.

In particular, a special form of the representation of the basic equations is established in which they assume a form, corresponding as it were to the independent movement of particles in an external field.

The actual interaction of the particles, as also in the method of secondary quantization, is here considered to be an operator structure of this [external] field.

The fundamental task of statistical physics is the study of large groups of interacting particles. Under these conditions either dynamical systems are considered which can be regarded as classical with a sufficient degree of approximation, or it is a question of systems, the quantum character of which is substantial. Specifically, the presence in the system of a large, in the limit, of an infinite number of particles, communicates to the problems of statistical physics their typical structure. It is of interest to note, that problems of a similar mathematical type are being arrived at in modern quantum field theory, in particular, in quantum electrodynamics.

At first glance it may seem, that the definite mathematical analogies which are being noted here have a purely formal character and do not rest on a more profound physical basis.

Actually, in the typical problems of quantum field theory we have to do with the study of the behavior either of a single particle, let us say, in an external field, or with the study of the interaction processes of two, in general, of a comparatively small number of particles. One would therefore think, that the presence of a characteristic peculiarity of the problems of statistical physics -- of a large group of particles or of a discrete medium, in which each individual particle moves, is completely absent.

In reasoning thus, we, however, lose sight of the fact that, according to the concepts of modern quantum field theory, each particle constantly interacts with a vacuum as with its own kind of quantum medium.

This aspect began to be noted immediately after the appearance of the Dirac theory, confirmed by the discovery of positrons which followed after it.

A vacuum then began to be spoken of as an aggregate of an infinite number of particles, filling the negative energy quantum states.

In spite of a certain paradoxicality and, in essence, if one can so express oneself, of the "pseudophysicity" of such a concept of the vacuum, it to a certain extent correctly indicated that a vacuum is not empty space in the ordinary popular understanding of the word "empty", but has definite properties of a discrete quantum medium.

The role of the vacuum has grown, particularly after the vigorous development of quantum field theory in its modern form which began in the last decade.

After such brilliant experiments as those pertaining to the anomalous magnetic moment of the electron and pertaining to the displacement of atomic levels, the reality of the effects of interaction of particles with a vacuum has now obtained universal recognition.

If one attempts to approach the problems of the statistical physics of quantum systems and the problems of quantum field theory from the point of view of the formal mathematical apparatus being used for their solution, we immediately note, that for them the possibility of the use of the method of secondary quantization appears to be general.

Under these conditions the operators of the generation and annihilation of particles $\Psi^{(+)}(\vec{r}),\ \Psi^{(-)}(\vec{r}),$ linked between themselves by definite commutation relations, are introduced, for example:

$$\Psi^{(-)}(\vec{r})\,\Psi^{(+)}(\vec{r}\,')-\Psi^{(+)}(\vec{r}\,')\,\Psi^{(-)}(\vec{r})=\delta(\vec{r}-\vec{r}\,'), \tag{1}$$

and the energy operator of the system and other dynamic quantities are expressed through them.

The wave function in the secondary quantization representation, designated sometimes as the amplitude of the state (the state-vector of American authors), is ordinarily considered as a function of the occupation numbers of the quantum levels of a single particle:

$$\Phi = \Phi(\ldots\ldots n_k \ldots\ldots).$$

One can also more clearly associate the wave functions of secondary quantization with the wave functions of the ordinary configurational representation.

Let, actually, Φ_0 be the amplitude for the state of the vacuum.

It is then evident, that the integral

$$\Phi(t) = \int \varphi(t, \vec{r}_1, \ldots \vec{r}_n) \Psi^{(+)}(\vec{r}_1) \ldots \Psi^{(+)}(\vec{r}_n) d\vec{r}_1 \ldots d\vec{r}_n \Phi_0 \qquad (2)$$

gives an expression for the amplitude of the state in which there are exactly n particles, c -- the function $\varphi(t, \vec{r}_1, \ldots \vec{r}_n)$ — here will be their wave function in the configurational representation.

Consequently, we shall obtain the amplitude of the state [state-vector] in the general case, in which the number of particles was not fixed, by virtue of the fundamental principles of quantum mechanics via the superposition of the amplitude of the state of type (2):

$$\Phi(t) = \sum_n \frac{1}{n!} \int \varphi(t, \vec{r}_1, \ldots \vec{r}_n) \Psi^{(+)}(\vec{r}_1) \ldots \Psi^{(+)}(\vec{r}_n) d\vec{r}_1 \ldots d\vec{r}_n \Phi_0. \qquad (3)$$

It thus turns out, that the amplitude of the state is completely characterized by a series

$$\ldots \varphi(t, \vec{r}_1, \ldots \vec{r}_n) \ldots \qquad (4)$$

of wave functions of the ordinary type.

This important fact was established by V. A. Fok. Fok also introduced a very interesting functional representation.

Let the field being considered correspond to particles with a Bose statistics, with commutation relations of type (1).

We shall then construct the functional:

[(5); see next page]

$$\Phi_t(U) = \sum_n \frac{1}{n!} \int \varphi(t, \vec{r}_1, \ldots \vec{r}_n) U(\vec{r}_1) \cdot U(\vec{r}_n) d\vec{r}_1 \ldots d\vec{r}_n \tag{5}$$

in some class of arbitrary functions of $U(\vec{r})$.

Inasmuch as all wave functions of series (4) are expressed through variational derivatives of this functional:

$$\varphi(t, \vec{r}_1, \ldots \vec{r}_n) = \left\{ \frac{\delta^n \Phi_t(U)}{\delta U(\vec{r}_1) \ldots \delta U(\vec{r}_n)} \right\} U \equiv 0,$$

we see that the amplitude of the state [state-vector] is completely characterized by the form $\Phi_t(U)$.

It is of interest to note, that in the Schrödinger equation

$$i \frac{\partial \Phi_t(U)}{\partial t} = H \Phi_t(U),$$

determining the time evolution of the form of the Fok functional, the H expression is obtained from the corresponding H expression of the method of secondary quantization by replacement of the generation operators $\Psi^{(t)}(\vec{r})$ by $U(\vec{r})$, and the annihilation operators $\Psi^{(-)}(\vec{r})$ by $\frac{\delta}{\delta U(\vec{r})}$.

Such "rules of transition" from the representation of secondary quantization to the representation of Fok are, in essence, perfectly evident.

Actually, in the secondary quantization representations neither one or another "realization" of the fundamental field operator functions has significance.

It is only essential that the recognized commutation relations be retained between them.

For multiplication operations on $U(\vec{r})$ and operations of variational differentiation $\frac{\delta}{\delta U(\vec{r})}$ there also is a commutation relation (1)

$$\frac{\delta}{\delta U(\vec{r})} U(\vec{r}') - U(\vec{r}') \frac{\delta}{\delta U(\vec{r})} = \delta(\vec{r} - \vec{r}'),$$

which is also [the case] for

$$\Psi^{(+)}(\vec{r}), \quad \Psi^{(-)}(\vec{r}).$$

Thus, in the Fok representation the Schrödinger equation will be an equation with variational derivatives.

If in this equation one now places a formal expansion of the functional $\Phi_t(U)$ in a power series of U and in the usual way compares the coefficients for identical powers in it, one obtains, generally speaking, an infinite series of equations for the determination of the wave functions (2).

In precisely such an "expanded" form the method of Fok has also obtained a wide application in subsequent investigations (let us mention, for example, the works pertaining to the approximate method of Tamm, pertaining to the approximate method of the intermediate bond of Tomonaga, the works of Yu. M. Shirokov, etc.).

Sufficient attention has not been devoted to equations in variational derivatives and to the functional $\Phi_t(U)$ determined by them, whereas their indirect study is technically much more convenient.

In this connection let us turn, for example, to the theory of the series of special functions of mathematical physics, where their study is substantially facilitated with the aid of the introduction of "generating functions", standing in exactly the same relation toward the special functions being studied there which is found between the functional of Fok and the series of wave functions (2).

Equations in variational derivatives have begun to attract ever greater attention to themselves only in a very recent period, after the appearance of the works of Ju. Schwinger pertaining to the theory of Green's functions.

Let us have a field of fermions Ψ, interacting with a field of bosons φ. In accordance with the fundamental idea of Schwinger let us add to the Lagrangian of the interaction

$$L(x) = g\varphi(x)\overline{\Psi}(x)\Gamma\Psi(x), \qquad x = (t, \vec{r}),$$

the term $\varphi(x)J(x)$, which can be interpreted as expressing the influence of the classical source J(x) of the boson field. Let us introduce the Green's function for fermions G(x, y) and the Green's function for bosons $\mathfrak{G}(x, y)$, letting:

$$\begin{aligned} G(x, y) &= i\frac{\langle T(\Psi(x)\overline{\Psi}(y)S)\rangle_0}{\langle S\rangle_0} \\ \mathfrak{G}(x, y) &= \frac{\delta}{\delta J(y)}\frac{\langle T(\varphi(x)S)\rangle_0}{\langle S\rangle_0} \\ S &= T(e^{i\int L(x/S)\,dx}). \end{aligned} \tag{6}$$

These Green's functions, or Greenians (as D. D. Ivanenko proposes to call them), evidently, are functionals of J(x) as of an arbitrary c-function.

The fundamental physical characteristics for the processes of interaction of one real fermion with one or several real bosons are directly expressed through their values and the values of their variational derivatives along J(x) at J = 0.

The case of the simultaneous presence of several real fermions also yields to description with the aid of Green's functions of an analogous structure which, it is true, is somewhat more complex.

Schwinger obtained for the introduced Green's functions (6) equations in variational derivatives of the type:

$$\left\{ i\gamma\partial - \mu - g\left(\Phi(x) - i\frac{\delta}{\delta J(x)}\right)\right\} G(x, y) = \delta(x - y). \tag{7}$$

$$(\Box + m^2)\,\mathfrak{G}(x, y) - ig\Gamma\frac{\delta}{\delta J(y)}G(x, y) = \delta(x - y);\ \frac{\delta\Phi(x)}{\delta J(y)} = \mathfrak{G}(x, y),$$

It is of interest to point out, that they can be constructed via the summation of ordinary expansions of S-matrices.

In this connection the hypothesis is advanced, that on the basis of the Schwinger theory the problem of the elimination of [tending to] infinity can be resolved without resort to the standard theory of perturbations and that a way will thereby be opened toward the construction of approximations of another kind, for example, belonging to the "intermediate" or "strong" coupling type.

In this direction, Edwards, Landau with coworkers, and others have already contributed a number of investigations having good prospects.

In connection with the form of equations (7), let us further note that the terms entering into them with variational derivatives correspond to a radiation correction or, which is the same, to an interaction with a vacuum.

If one rejects these terms, then one obtains the well known equations for free fermions in an external field.

It is not difficult to reconstruct the Schwinger equations in variational derivatives and bring them to the form of an infinite series of equations linked one after the other for the determination of the succession of functions

$$\ldots G_n(x, y, \xi_1, \ldots \xi_n) \ldots,$$

corresponding to the amplitudes of scattering of a single real fermion in the presence of interaction with n real bosons.

For this one has but to place in equation (7), as also in the method of Fok, a formal expansion of G(x, y) in the power series of J [involving J's]:

$$G(x, y) = G(x, y/J) = \sum_{(n)} \frac{1}{n!}\int G_n(x, y, \xi_1, \ldots \xi_n)\, J(\xi_1)\ldots J(\xi_n)\, d\xi_1 \ldots d\xi_n.$$

Generally speaking, it is more convenient to have to deal not with G_n, but with the so-called nodal parts $\Gamma_n(x, y, \xi_1, \ldots \xi_n)$:

$$\Gamma_n(x, y, \xi_1, \ldots \xi_n) = \left\{ \frac{\delta^n G^{-1}(x, y/J)}{\delta J(\xi_1) \ldots \delta J(\xi_n)} \right\}_{J=0}.$$

In order to obtain a chain of equations, for such Γ_n it is sufficient to pass in equations (7) from $G(x, y)$ to $G^{-1}(x, y)$ and to avail oneself of the expansion:

$$G^{-1}(x, y) = \sum_{(n)} \frac{1}{n!} \int \Gamma_n(x, y, \xi_1, \ldots \xi_n) J(\xi_1) \ldots J(\xi_n) d\xi_1 \ldots d\xi_n.$$

A system of equations constructed in such a fashion, just as other systems similar to it, is infinite.

One can attempt to break it, devising some approximate formula. expressing the given Γ_n via the preceding $\Gamma_0, \Gamma_1, \ldots \Gamma_{n-1}$.

Then the system of the first n-equations of the chain is rendered closed. Landau, Abrikosov, Khalatnikov, and also M. Neĭman have entered upon this path in their investigations.

It should be underlined, that the chains of equations, obtained by the method of Schwinger, are different in principle from the corresponding chains of the method of Fok in their four-dimensionality.

But just as the method of Fok also, the method of Schwinger is closely linked with the representation of secondary quantization.

Attention has not been directed to this important circumstance up to this time, although here we can arrive at a novel treatment of the equations (7) which has good prospects.

Let us agree to consider

$$J(x), \quad \pi(x) = \frac{\delta}{\delta J(x)}$$

as operators, linked by the commutation relations:

$$[\pi(y); \ J(x)] = \delta(x - y). \tag{8}$$

Then the equations (7) assume a "secondary-quantized" form:

$$\begin{aligned} \{i\gamma\partial - \mu - g\Gamma(\Phi(x) - i\pi(x))\} G(x, y) &= \delta(x - y), \\ (\Box + m^2) \mathfrak{G}(x, y) - ig\Gamma\pi(y) G(x, y) &= \delta(x - y) \end{aligned} \tag{9}$$

with the commutation relations:

$$\begin{aligned} [\Phi(x); \ J(y)] &= 0, \\ [\pi(y); \ \Phi(x)] &= \mathfrak{G}(x, y). \end{aligned} \tag{10}$$

It is of interest to note, that equations (9), (10) form a nonlinear system, and, what is more, the <u>four-dimensional</u> commutation relations (10) are not fixed, but are determined by the entire aggregate of equations.

Here, apparently, we naturally approach the nonlinear theory long since sought for in principle.

Let us note yet again that the given form of the representation of the Schwinger equations is very convenient for obtaining a different kind of approximate solutions.

Let us consider, for example, the "single-particle" approximation recently proposed by us in which, speaking in the language of the Feynman diagrams, we neglect the closed loops and consider the fermion line to be connected.

It is proposed by the same token, that the boson Green's function $\mathfrak{G}(x, y)$ can be approximately replaced by its value for a free field, i.e., by the ordinary causal function $D^c(x - y)$.

In such a "single-particle approximation" the system (9) is rendered linear, being reduced to the single equation

$$\{i\gamma\partial - \mu - \partial\Gamma\vartheta(x)\} G(x, y) = \delta(x - y), \tag{11}$$

in which:

$$\vartheta(x) = \Phi(x) - i\pi(x) = \int D^c(x - y) J(y)\, dy - i\pi(x).$$

From (10) we find:

$$[\pi(y);\ \Phi(x)] = D^c(x - y) \tag{11'}$$

and consequently, thanks to the symmetry of $D^c(x - y)$ we shall have:

$$[\vartheta(x);\ \vartheta(y)] = 0. \tag{12}$$

As is evident, we arrive at a problem, perfectly similar in form to the problem of the movement of a single secondarily nonquantized particle in a quantum boson field, which was already being studied long before the appearance of the current conceptions of quantum field theory. It is to the point to state, that up to the present time it [the problem] is still being considered in connection with work with the methods of strong or intermediate coupling, inasmuch as these last have still not been successfully applied toward modern variants of the field theory.

The analogy is rendered still more direct, if one passes from the Green's function $G(x, y)$ to the corresponding wave function $\varphi(x)$. We then obtain the single-particle equation:

$$(i\gamma\partial - \mu - g\Gamma\vartheta(x))\,\varphi(x) = 0. \tag{13}$$

The sole difference, it is true, a very substantial one, of our single particle approximation from the usual one consists in the form of the commutation relations (11').

If we had wished in our scheme to completely pass to the usual single-particle problem, we would have had to replace in (11') the causal function $D^c(x - y)$ by

$$- D^{(+)}(x-y).$$

Actually, in place of (12) we would then obtain the commutation relations:

$$[\vartheta(x);\ \vartheta(y)] = -iD^{(+)}(x-y) - iD^{(-)}(x-y) = -iD(x-y)$$

with the ordinary Pauli function.

From the physical point of view the difference which has been mentioned in the commutation relations is occasioned by the fact that in our approximation we do not completely neglect the virtual generation of pairs, since the fermion line on the Feynman diagrams can go both "forward", and also "backward" in time.

This difference does not, however, prevent the possibility of transfer of the methods, worked out for the ordinary single particle approximation, to our scheme.

For this purpose it is convenient to exclude from the corresponding equations (11), (13) the coordinates of the fermions.

Availing ourselves of the properties of translational invariance, we put

$$G(x, y/J) = (2\pi)^{-4} \int e^{i(k.y-x)},\quad G(k/T_x J)\,dk, \tag{14}$$
$$\varphi(x/J) = e^{-i(k.x)}\ \varphi(T_x J),$$

where T_x represents the translation:

$$J(\xi) \rightarrow J(\xi + x).$$

We then obtain:

$$\left\{\gamma\left(k + i\int pJ(p)\,\pi(p)\,dp\right) - \mu - g\Gamma(2\pi)^{-2}\left(\int D^c(p)\,J(p)\,dp - \right.\right.$$
$$\left.\left. - i\int \pi(p)\,dp\right)\right\} G(k/J) = 1 \tag{15}$$

and

$$\left\{\gamma\left(k + i\int pJ(p)\,\pi(p)\,dp\right) - \mu - g\Gamma(2\pi)^{-2}\left(\int D^c(p)\,J(p)\,dp - \right.\right.$$
$$\left.\left. - i\int \pi(p)\,dp\right)\right\} \varphi(J) = 0, \tag{16}$$

and also

$$\pi k = \frac{\delta}{\delta J(k)}.$$

The last equation (16) belongs to the eigenvalue problem: equation (15) determines the Fourier-components of the Green's function. It is characteristic, that these components are not confused with each other and are determined individually for each "value" of the four-dimensional vector k.

Let us note in conclusion, that one can directly apply the Bloch-Nordsieck method, the methods of strong and intermediate coupling, to the equations obtained in the form of (15, (16).

Up to this time we have spoken of equations with variational derivatives in connection with questions of quantum theory.

The sphere of their effectiveness is, however, considerably more extensive.

Let us consider, for example, the problem of the statistical equilibrium of a classical system of identical particles with interaction of a binary type, a problem, one would think, very far from quantum theory.

We shall base ourselves here on the ordinary canonical Gibbs distribution:

$$D(\vec{r}_1, \dots \vec{r}_N) = Q_N^{-1} \exp\left\{-\frac{U_N}{\theta}\right\},$$

where

$$U = \sum \Phi(r_{ij}),$$

while Q_N is the configurational integral.

As was shown in our work of 1945, the corresponding functions of the distribution of complex particles

$$F_s(\vec{r}_1, \dots \vec{r}_s) \tag{17}$$

are conveniently studied with the aid of a "generating functional"

$$L(U) = \int \left\{\Pi\left(1 + \frac{V}{N} U(\vec{r}_j)\right)\right\} D(\vec{r}_1, \dots \vec{r}_N)\, d\vec{r}_1 \dots d\vec{r}_N. \tag{18}$$

These distribution functions are expressed via variational derivatives:

$$F_s(r_1, \dots r_s) = \frac{\delta^s L(U)}{\delta U(\vec{r}_1) \dots \delta U(\vec{r}_s)}.$$

An equation in variational derivatives was obtained for L(U) which is also directly transformed into the form of a chain of equations for determination of the functions (17).

Let us note, incidentally, that for those cases when it can not be a question of any small parameter (as, for example, in the study of the liquid state), an approximation expressing F_n via the preceding -- [e.g.], F_3 via F_2 -- has been proposed. By the same token, a break in the chain and the possibility of the factual solution of the equations being obtained is achieved. At the present time such an approximation is being successfully developed by Fisher.

As is evident, in the classical theory of the statistical equilibrium of condensed systems the question of the structure of approximate solutions has a great similarity to the analogous question of quantum field theory for the case in which the interaction with a vacuum is neither strong nor weak.

The functional L(U) introduced by us does not have a direct physical interpretation.

It is easy, however, to construct a functional which is very similar, but which more "graphically physical".

We avail ourselves for this of the ideas of the energizing [inclusion] of an external field, in exactly the same way as in the Schwinger theory of the Green's functions.

Let the given aggregate of particles be found in an external field. Then the free energy

$$F = -\Theta \ln \int \exp\left\{-\frac{1}{\Theta} U_N - \frac{1}{\Theta} \sum \varphi(\vec{r}_j)\right\} dr_1 \ldots dr_N \tag{19}$$

can be considered as a functional of the external field:

$$F = F(\varphi).$$

It is perfectly clear, that via this variational derivative one can also express all the distribution functions of complexes of particles in terms of ? .

Let us note, in particular, that

$$\frac{\delta F(\varphi)}{\delta\varphi(\vec{r})} = F_1(\vec{r}/\varphi)$$

and

$$F_2(\vec{r}, \vec{r}'/\varphi) = \frac{N}{N-1} F_1(\vec{r}/\varphi) F_1(\vec{r}'/\varphi) - \frac{V}{N-1} F_1(\vec{r}/\varphi)\,\delta(\vec{r}-\vec{r}') -$$
$$- \frac{V\Theta}{N-1} \frac{\delta F_1(\vec{r}/\varphi)}{\delta\varphi(\vec{r})}. \tag{20}$$

Here F_1, F_2 are distribution functions in the presence of an external field.

Let us now consider the equation for F_1.

We have:

$$\frac{\partial F_1(\vec{r}/\varphi)}{\partial r^\alpha} + \frac{1}{\Theta} \frac{\partial\varphi(\vec{r})}{\partial r^\alpha} F_1(\vec{r}/\varphi) +$$
$$+ \frac{N-1}{V\Theta} \int \frac{\partial\Phi(r_{12})}{\partial r^\alpha} F_2(\vec{r}, \vec{r}'/\varphi)\, d\vec{r}' = 0.$$

Substituting here in place of F_2 its expression (20), we obtain the following equation in variational derivatives:

$$\frac{\partial F_1(\vec{r})}{\partial r^\alpha} + \frac{1}{\Theta}\frac{\partial \vartheta(\vec{r})}{\partial r^\alpha} F_1(\vec{r}) = 0,$$

$$\vartheta(\vec{r}) = \int \Phi(r_{12}) \left\{ \frac{N}{V} F_1(\vec{r}') - \delta(\vec{r} - \vec{r}') - \Theta \frac{\delta}{\delta\varphi(\vec{r}')} \right\} d\vec{r}' + \varphi(\vec{r}) \qquad (21)$$

for the function $F_1(\vec{r}) = F_1(\vec{r}/\varphi)$, representing the ordinary density of the number of particles in the presence of the external field.

We here detect a very interesting fact -- this density $F_1(\vec{r})$ satisfies exactly just such an equation in form, as also in the case when every particle moves independently of the others in the fixed external field $\vartheta(\vec{r})$.

The presence of interparticle interaction manifests itself only in the fact that $\vartheta(\vec{r})$ is not an ordinary c-function, but an operator function, containing variational differentiation.

A situation is created, completely analogous to the situation in the many-body problem of quantum mechanics, when, using the method of secondary quantization, we obtain just such an equation in form as for a single individual particle, but only with an "external field" having an operator character.

Neglecting the operator character of $\vartheta(\vec{r})$, we arrive at the equation of a "self-consistent field", in the classical case which has been considered by A. A. Vlasov.

The considerations set forth above can also be utilized in the study of non-equilibrium states.

Thus, the operator method of the type of secondary quantization, the method of equations with variational derivatives, is applicable not only to quantum systems, but also to classical systems.

Its effectiveness is associated with the presence of a large group of particles (it makes no difference, [whether they are] real or virtual), and not with the quantum or classical nature of the dynamical systems being studied.

In establishing a bond between problems from what one would have thought were the divisions of theoretical physics the most remote from each other, this method facilitates the transfer of methods for the solution of such problems from one field to another and, undoubtedly, has wide prospects of further development in subsequent investigations.

THE METHOD OF FUNCTIONALS IN THE QUANTUM THEORY OF FIELDS

YU. V. NOVOZHILOV AND A. V. TULUB

I. The Method of Functionals in the Quantum Theory of Fields

§ 1. The quantum theory of fields and functionals

It is known that a field is characterized, from the classical point of view, by one or several functions of coordinates $\varphi(x)$, the field potentials, which satisfy the wave equation. A field can be considered as a mechanical system having an infinite number of degrees of freedom. Indeed, in order to determine classically the state of a system of a finite number of degrees of freedom n we have to assign to the system n, independent coordinates q_l, and n conjugate momenta p_l $(l=1, 2 \ldots n)$. Hence the state of a field will be known if for each of the points $x_1, x_2, x_3 \ldots$ of space we know the potential and the derivative with respect to time $\partial\varphi/\partial t$, i.e., if we know the infinite totality of quantities $\varphi(x_1), \frac{\partial\varphi(x_1)}{\partial t}; \varphi(x_2), \frac{\partial\varphi(x_2)}{\partial t}; \ldots$

In the case of a system with a finite number of degrees of freedom, we have to deal with functions of $f(q_1 \ldots q_n\, p_1 \ldots p_n)$ of the coordinates q_l and of the canonically conjugate momenta p_l (energy, momentum, etc). Hence in

---

* § 3 and § 4.2 are written by A.V. Tulub and the rest by Yu. V. Novozhilov.

the case of a field, the energy and the other field quantities will depend on the infinitely many "coordinates" $\varphi(x_i)$, where x_i runs over all the points of space, and on the conjugate "momenta". Thus we have to consider in a field theory, functions of infinitely many variables $\varphi(x_1), \ldots, \varphi(x_i)$ or functions of the function $\varphi(x)$, i. e., functionals of the function $\varphi(x)$*.

The dependence of the functional upon its arguments will be denoted by braces: $F\{\varphi(x)\}$ or $F\{\varphi\}$ is a functional of $\varphi(x)$. The simplest example of a functional is the integral $F\{\varphi\} = \int f(x)\varphi(x)\,dx$.

Let us consider the variation $\delta F\{\varphi\}$ of an arbitrary functional $F\{\varphi\}$, caused by the variation of the function $\varphi(x)$ at the point x': $\delta F\{\varphi\} = F\{\varphi(x) + \lambda\delta(x-x')\} - F\{\varphi\}$. The quantity

$$\frac{\delta F\{\varphi\}}{\delta\varphi(x')} = \lim_{\lambda\to 0}\frac{1}{\lambda}\,\delta F\{\varphi\}. \tag{1. 1}$$

is called the functional derivative of $F\{\varphi\}$ with respect to $\varphi(x')$

Instead of (1. 1) we may also write

$$\delta F\{\varphi\} = \int \frac{\delta F\{\varphi\}}{\delta\varphi(x)}\,\delta\varphi(x)\,dx. \tag{1. 2}$$

Hence we obtain, by putting

$$F\{\varphi\} = \varphi(x) = \int \delta(x-x')\,\varphi(x')\,dx',$$

that

$$\frac{\delta\varphi(x)}{\delta\varphi(x')} = \delta(x-x'). \tag{1. 3}$$

If $F\{\varphi\}$ is a functional of the function $\varphi(x)$, which is itself a functional of the function $\gamma(x)$, then the functional derivative of F with respect to $\gamma(x')$ is equal to

$$\frac{\delta F\{\varphi\{\gamma\}\}}{\delta\gamma(x')} = \int \frac{\delta\varphi(x)}{\delta\gamma(x')}\,\frac{\delta F\{\varphi\{\gamma\}\}}{\delta\varphi(x)}\,dx. \tag{1. 4}$$

The definition of the functional integral will be considered by us later on**.

It is well known that in the quantum mechanics of a system with a finite number degrees of freedom n, the fundamental commutation relations be-

----

* [Translator's note: A functional, strictly speaking, is a mapping of a set of point sets onto a point set. A function is a mapping of a point set onto a point set. The above definition implies a functional in the strict sense, and the term "function of a function", although used in literature, is apt to be misleading. See Volterra,, "The Theory of Functionals", Blackie and Sons, London, 1930.]

* Some mathematical problems relating to the formulation of the quantum theory of fields in terms of functionals are considered in Friedrichs' book /11/ and in the appendix to Symanzik's article /12/.

tween the operators of the coordinates q_l and of the momenta $p_{l'}$ have the form

$$[\hat{p}_{l'}, \hat{q}_l] = \hat{p}_{l'}\hat{q}_l - \hat{q}_l\hat{p}_{l'} = -i\delta_{ll'}. \qquad (1.5)$$

Since all the coordinate operators q_l commute among themselves, we can describe the system by means of the wave function $\Psi(q_1,\ldots,q_n)$, , in relation to which $\hat{q}_l$ is a number (the operation $\hat{q}_l$ consists of multiplication by q_l). From (1.5) follows then, as is generally known, that $\hat{p}_l$ is the differential operator:

$$\hat{p}_l = -i\frac{\partial}{\partial q_l}.$$

If we now pass to an infinite number of degrees of freedom by letting n tend to infinity, then the index l will run over a continuous set of values. The infinite sets of coordinates $\hat{q}_l$ and momenta $\hat{p}_l$ thus obtained, may be considered as functions $q(l)$ and $p(l)$, depending on the parameter l In that limiting case the commutation relation (1.5) ought to be written in the form

$$[\hat{p}(l), \hat{q}(l')] = -i\delta(l-l'), \qquad (1.6)$$

where on the right-hand side we replaced $\delta_{ll'}$ by $\delta(l-l')$, since l is a continuous variable. The wave function Ψ will depend on the infinitely many coordinates q_l or on the function $q(l)$, i.e., Ψ will be a functional of $q(l)$: $\Psi = \Psi\{q(l)\}$.

In the quantum theory of fields the fundamental commutation relation for fields with integral spin has precisely the form (1.6). The operator $\varphi(x)$ of a meson field (the meson "potential") depends parametrically on the space–time coordinate x. If the operators $\varphi(x')$ and $\partial\varphi(x)/\partial x_0$ are taken at the same instant, their commutator is

$$\left[\frac{\partial\varphi(x)}{\partial x_0}, \varphi(x')\right] = -i\delta^3(x-x'), \quad x_0 = x_0'. \qquad (1.7)$$

Since the operators $\varphi(x)$ and $\varphi(x')$ commute with each other for different points x and x' at equal times $x_0 = x_0'$, we may choose as $\varphi(x)$ (with fixed time) the coordinate function $\hat{q}(l)$, i.e., consider $\varphi(x)$ as the operator of multiplication by the function $\varphi'(x)$, and describe the field by means of the functional $\Omega\{\varphi'\}$ of the space function $\varphi'(x)$. It then follows from the commutation relation (1.7) that the operator $\partial\varphi/\partial x_0$ plays the role of the momentum function $\hat{p}(l)$. Relation (1.7) will be satisfied according to (1. 3) if we put

$$\frac{\partial \varphi(x)}{\partial x_0} \Omega\{\varphi'\} = -i \frac{\delta}{\delta \varphi'(x)} \Omega\{\varphi'\}. \tag{1.8}$$

Thus the functional formulation in the quantum theory of fields is the natural consequence of the fact that a field has an infinite number of degrees of freedom.

The concept of the generating functional plays an important part in the quantum theory of fields. We assume that the functional $F\{\varphi\}$ of the function $\varphi(x)$ can be expanded in a functional power series

$$F\{\varphi\} = \sum_n F_n\{\varphi\}; \tag{1.9}$$

$$F_n\{\varphi\} = (n!)^{-1/2} \int f_n(x_1 \ldots x_n)\, \varphi(x_1) \ldots \varphi(x_n)\, dx_1 \ldots dx_n. \tag{1.10}$$

Then $F\{\varphi\}$ is the generating functional for the functions $f_n(x_1 \ldots x_n)$ which are symmetrical in the variables $x_1 \ldots x_n$. The concept of the generating functional is a generalization of the concept of the generating function In the simple case when the variable x can assume only one value $x = a$, we obtain from (1.10), by putting $\varphi(x) = \zeta\delta(x-a)$:

$$F\{\varphi(a)\} \equiv F(\zeta) = \sum_{n=0}^{\infty} \frac{1}{\sqrt{n!}} f_n \zeta^n$$

i.e., the generating function for the coefficients f_n.

There is a one to one correspondence between the functional $F_n\{\varphi\}$ and the function $f_n(x_1 \ldots x_n)$ which permits the use of the functional $F_n\{\varphi\}$ instead of the function f_n. This means that instead of the infinite set of functions $f_n(x_1 \ldots x_n)$, $n = 0,1,2,\ldots$, contained in (1. 10), we may deal with one quantity - the functional $F\{\varphi\}$.

The method of the generating functional may be extended to the case of functions $f_n(x_1 \ldots x_n)$ which are antisymmetric in the variables $x_1 \ldots x_n$. In this case the quantities $\varphi(x)$ contained in the integral $F_n\{\varphi\}$ (formula 1,10) cannot be functions. Indeed, if $\varphi(x)$ is a function and $f_n(x_1 \ldots x_n)$ is antisymmetric in the variables $x_1 \ldots x_n$, then the integral F_n vanishes. In order that the integral F_n should remain unchanged for a permutation of two variables x_i and x_k, the quantities $\varphi(x)$ ought to anticommute: $\varphi(x_l)\varphi(x_k)+\varphi(x_k)\varphi(x_l)=0$. The meaning of the anticommuting quantities φ will be explained in § 2.

The method of the generating functional is closely connected with the particle aspect of the quantum theory of fields. The interpretation of the transitions and states of the field is based on the concept of the particle aspect of the field. As long as this concept holds, we have to consider, as the basic quantities which describe the field, the wave functions $f_n(x_1 \ldots x_n)$, belonging to a given number of particles n and possessing the required symmetry properties with respect to the variables $x_1 \ldots x_n$. If the number of particles is variable, then the state on the transitions of the field ought to be described, in the general case, by the totality of functions $f_n(x_1 \ldots x_n)$[13], $n=0, 1, 2, \ldots$:

$$\begin{array}{l} f_0, \\ f_1(x_1), \\ f_2(x_1, x_2), \\ \ldots\ldots\ldots\ldots\ldots \\ f_n(x_1, x_2, \ldots, x_n). \\ \ldots\ldots\ldots\ldots\ldots \end{array}$$

Instead of the totality of functions $f_n(x_1 \ldots x_n)$ we may consider a generating functional of the type (1.10).

Thus a functional may describe the state of a field or the transition processes only if it is a generating functional with respect to some functions $f_n(x_1 \ldots x_n)$. The simplest functions which depend on n particles are the probability amplitudes $\Psi_n(x_1 \ldots x_n)$, the square of their absolute values $|\Psi_n|^2$ giving the probability density of the particles being in the states $x_1 \ldots x_n$. The generating functional for the probability amplitudes is the "Fock functional". We know at present, in addition to the set of probability amplitudes $\Psi_n(x_1 \ldots x_n)$ other functions (four dimensional wave functions, Green's functions), which may be used for the description of the states of the field and of its transitions; the probability amplitude treatment is close to non-relativistic quantum mechanics, whereas in explicit definitions of the four-dimensional wave functions and Green's functions, the requirement of relativistic invariance of the theory is taken into account. These relativistic functions are connected with Schwinger's functional of external sources.

The four-dimensional wave functions, Green's functions and other relativistic functions do not have a simple meaning like the probability amplitudes.

The method of functionals will be presented for a nucleon field interacting with a neutral pseudoscalar meson field. We denote the operators of the free nucleon field by $\psi_\alpha(x)$ and $\overline{\psi}_\beta(x)$ $(\overline{\psi}=\psi^*\gamma_4)$, the operator of the free meson field by $\varphi(x)$ the spin variables α, β will be lumped together with the coordinates in the same letter, e. g., $\psi(x)$, $\overline{\psi}(y)$, and integration over the coordinates x, y will include summation over the remaining variables.

It is necessary to introduce, for subsequent use, the operators of creation and annihilation. $f^{(k)}_{(x)}$ are a system of positive frequency solutions of the Klein-Fock equation $(\Box-\mu^2)f^{(k)}_{(x)}=0$, which are orthogonal and normalized so that

$$-i\int d\sigma_\mu f^{(j)}(x)\frac{\overleftrightarrow{\partial} f^{(k)}_{(x)}}{\partial x_\mu}=\delta_{jk} \tag{1.11}$$

($d\sigma_\mu$ being the four-vector element of the space-like hypersurface), then the annihilation operator c_k for the meson field $\varphi(x)$ is defined by formula

$$c_k=\frac{1}{i}\int d\sigma_\mu\varphi(x)\frac{\overleftrightarrow{\partial}\overline{f}^{(k)}_{(x)}}{\partial x_\mu}. \tag{1.12}$$

In (1.11) and (1.12) we have used the notation

$$A\frac{\overleftrightarrow{\partial}}{\partial x_\mu}B=A\frac{\partial B}{\partial x_\mu}-\frac{\partial A}{\partial x_\mu}B. \tag{1.13}$$

Let $u^{(k)}$ and $\overline{v}^{(k)}$ be systems of positive-frequency solutions of Dirac's equation $(u^{(k)})$ and the [charge-]conjugate equation $(\overline{v}^{(k)})$, which are orthogonal and normalized $(v=v^*\gamma_4)$

$$\left.\begin{aligned}&\int\overline{u}^{(k)}\gamma_\mu u^{(j)}d\sigma_\mu=\delta_{kj},\\&\int\overline{v}^{(k)}\gamma_\mu v^{(j)}d\sigma_\mu=\delta_{kj};\quad\int\overline{u}^{(k)}\gamma_\mu v^{(j)}d\sigma_\mu=0.\end{aligned}\right\} \tag{1.14}$$

Then the annihilation operators a_k and b_k for the nucleon field are defined by the equations

$$\left.\begin{aligned}a_k&=\int\overline{u}^{(k)}_{(x)}\gamma_\mu\psi(x)d\sigma_\mu,\\b_k&=\int\overline{\psi}(x)\gamma_\mu v^{(k)}(x)d\sigma_\mu.\end{aligned}\right\} \tag{1.15}$$

The creation operators a_k^+, b_k^+, c_k^+ are Hermitian conjugates of the annihilation operators a_k, b_k, c_k. If we consider the field at a moment

x_0 and the states of the particles differ in the value of the momentum, then the annihilation and creation operators defined by formulas (1.12) and (1.15) are the coefficients of the corresponding Fourier expansions:

$$\varphi(x) = \frac{1}{(2\pi)^{3/2}} \int (2k_0)^{-1/2} [c(k) e^{ikx} + c^+(\mathbf{k}) e^{-ikx}] d^3k, \tag{1.16}$$

$$\psi(x) = \frac{1}{(2\pi)^{3/2}} \int [u(\mathbf{p}) a(\mathbf{p}) e^{ipx} + v(\mathbf{p}) b^+(\mathbf{p}) e^{-ipx}] d^3\mathbf{p}, \tag{1.17}$$

$$\bar{\psi}(x) = \frac{1}{2(\pi)^{3/2}} \int [\bar{u}(\mathbf{p}) a^+(\mathbf{p}) e^{-ipx} + \bar{v}(\mathbf{p}) b(\mathbf{p}) e^{ipx}] d^3p, \tag{1.18}$$

where $kx = (\mathbf{k}, \mathbf{x}) - k_0 x_0$; and u and v are Dirac spinors; in (1.17) and (1.18) integration over p includes summation over the polarizations.

The commutation relations between the creation operators a^+, b^+, c^+ and the annihilation operators a, b, c have the form

$$\{a^+(\mathbf{p}), a(\mathbf{p}')\} = \delta^3(p - p'), \tag{1.19}$$

$$\{b^+(\mathbf{p}), b(\mathbf{p}')\} = \delta^3(p - p'), \tag{1.20}$$

$$[c(\mathbf{p}), c^+(\mathbf{p}')] = \delta^3(p - p'). \tag{1.21}$$

§ 2. The method of Fock functionals

1. The essence of the method. We shall at first expound the essence of the method with a very simple example so as to avoid complications which arise through the introduction of functionals. Hence let us consider, for illustrating the method, the case where the mesons may be found only in one state (the same for all particles). We shall not consider for the time being, the nucleon field. Let Ψ_n be the probability amplitude of finding n mesons (which depends only on the number n) in the field. In the case of an undetermined number of mesons it is necessary to know, for a complete description of the field, the entire set of values of Ψ_n, $n = 0, 1, 2, \ldots, \infty$.

In the method of Fock functionals the state of the field is characterized, instead of the infinite set Ψ_n, by the single quantity:

$$\omega(\bar{c}) = \sum_{n=0}^{\infty} \frac{1}{\sqrt{n!}} \Psi_n \bar{c}^n = \sum_{n=0}^{\infty} \omega_n, \tag{2.1}$$

where $\bar{c}$ is an auxiliary constant. It is obvious that the assignment of ω_n uniquely determines Ψ_n and vice versa. Thus, in our very simple example,

ω is a function of $\bar{c}$, and also the generating function for the probability amplitudes Ψ_n.

In order to allow the calculation of the mean values of the field quantities by means of the single quantity $\omega(\bar{c})$, for the purpose of writing the equation for $\omega(\bar{c})$, it is necessary to determine the effect of the field operators on $\omega(\bar{c})$.

The creation operators $c+$ and annihilation operators c satisfy in our example the commutation relation

$$cc^+ - c^+c = 1, \tag{2.2}$$

and the number-of-particles operator $\hat{n}$ is

$$\hat{n} = c^+c. \tag{2.3}$$

We shall show that $\omega(\bar{c})$ is the wave function of the field in a representation where $c+$ is the operator which consists in multiplication by $\bar{c}$:

$$c^+\omega(\bar{c}) = \bar{c}\,\omega(\bar{c}). \tag{2.4}$$

If (2.4) holds, it follows from commutation relations (2.2) that c is the operator of differentiation with respect to $\bar{c}$:

$$c\omega(\bar{c}) = \frac{d}{d\bar{c}}\,\omega(\bar{c}). \tag{2.5}$$

We may now write the equation for the eigenfunctions λ_n of the number-of-particles operator $\hat{n}$ in the form

$$\bar{c}\frac{d}{d\bar{c}}\lambda_n(\bar{c}) = n\lambda_n(\bar{c}), \tag{2.6}$$

where n (which is a positive integer denotes the number of particles. The solution of (2.6) is a power function of $\bar{c}$:

$$\lambda_n = A_n\bar{c}^{\,n}, \tag{2.7}$$

where A_n is a normalizing factor determined by the condition that the scalar product of λ_n by itself should equal unity: $(\lambda_n, \lambda_n) = \lambda_n^* \lambda_n = 1$. Hence, we immediately obtain that $|A_0| = 1$ and, as a consequence, in this representation the normalized function of a state without particles $\lambda_0 = 1$. In view of this we may also represent λ_n in the form

$$\lambda_n = A_n (c+)^n \lambda_0.$$

The scalar product of the eigenfunctions λ_n and λ_m will be equal to

$$(\lambda_n, \lambda_m) = A_n^* A_m ((c+)^n \lambda_0, (c+)^m \lambda_0) =$$
$$= A_n^* A_m (\lambda_0, c^n (c+)^m \lambda_0) = A_n^* A_m \frac{d^n}{d\bar{c}^n} \bar{c}^m \Big|_{\bar{c}=0} =$$
$$= |A_n|^2 n! \, \delta_{nm}, \text{ or } A_n = (n!)^{-1/2}, \tag{2.8}$$

where we used (2.4), (2.5) and the fact that the operators c and $c+$ are Hermitian conjugates.

Thus, series (2.1) represents an expansion of the function $\omega(c)$ in terms of the eigenfunctions $\lambda_n(\bar{c})$ of the number-of-mesons operator, with coefficients Ψ_n Since the meaning of Ψ_n is known to us in advance (the probability amplitude of finding n mesons in the field), $\omega(\bar{c})$ is indeed the wave function of the field.

The scalar product of the wave functions ω and ω' is, by (2.1) and (2.8) equal to

$$(\omega', \omega) = \sum_n \Psi_n'^* \Psi_n, \tag{2.9}$$

and for the probability amplitudes Ψ_n we may write the expression

$$\Psi_n = (n!)^{-1/2} (\lambda_0, c^n \omega), \tag{2.10}$$

or

$$\Psi_n = (n!)^{-1/2} \frac{d^n}{d\bar{c}^n} \omega(\bar{c}) \Big|_{\bar{c}=0}. \tag{2.11}$$

By virtue of (2.10) we obtain for the generating functional $\tilde{\omega}$ of the functions $\tilde{\Psi}_n = \sqrt{n!}\,\Psi_n$ the equation

$$\tilde{\omega}(\bar{c}') = (\lambda_0, e^{c\bar{c}'} \omega). \tag{2.12}$$

In order to explain (2.12) it should be noted that, in a representation where $c+$ is the multiplication operator, the eigenfunction $\omega_{c'}$ of the annihilation operator c is determined by equation

$$\frac{d}{d\bar{c}} \omega_{c'}(\bar{c}) = c' \omega(\bar{c}),$$

hence

$$\omega_{c'} \sim \exp[c'c];$$

where c' is the eigenvalue of the operator c.

2. The generating functional for probability amplitudes. Let us pass to the real case. We shall denote the state of the meson by its momentum $\mathbf{p}$. The probability amplitude of finding in the field n mesons with momenta $\mathbf{p}_1, \mathbf{p}_2 \ldots \mathbf{p}_n$ will now be a function $\Psi_n(\mathbf{p}_1 \ldots \mathbf{p}_n)$ of variable mesons. In the general case, the state of the field will be characterized, as hitherto, by the totality of infinitely many amplitudes $\Psi_n(\mathbf{p}_1 \ldots \mathbf{p}_n)$, $n = 0, 1, 2 \ldots \infty$, each of which describes a system of n mesons in the momentum space. The functions $\Psi_n(\mathbf{p}_1 \ldots \mathbf{p}_n)$ are symmetrical in their variables.

Let us introduce the auxiliary function of a vector argument $\bar{c}(\mathbf{p})$ and let us consider, instead of the wave function $\Psi_n(\mathbf{p}_1 \ldots \mathbf{p}_n)$, the functional of the function $\bar{c}(\mathbf{p})$:

$$\Omega_n\{\bar{c}\} = \frac{1}{\sqrt{n!}} \int \Psi_n(\mathbf{p}_1 \ldots \mathbf{p}_n)\, \bar{c}(\mathbf{p}_1) \ldots \bar{c}(\mathbf{p}_n)\, d^3\mathbf{p}_1 \ldots d^3\mathbf{p}_n. \qquad (2.13)$$

Since the assignment of $\Omega_n\{\bar{c}\}$ uniquely determines Ψ_n and vice versa, the state of the field may be described by the generating functional:

$$\Omega\{\bar{c}\} = \sum_{n=0}^{\infty} \Omega_n\{\bar{c}\}. \qquad (2.14)$$

instead of the set of probability amplitudes Ψ_n. Let us explain the meaning of the functional $\Omega\{\bar{c}\}$. We shall see that, just as in the elementary example of § 2, 1, $\Omega\{\bar{c}\}$ is the field-state vector in a representation where the creation operator $c^+(\mathbf{p})$ is a multiplication operator. In order to convince ourselves that it is so, let us consider the state vector Φ in such a representation. Since the operator $c^+(\mathbf{p})$ now depends on the meson momentum as a parameter, its effect on the vector of state Φ consists in the multiplication of Φ by a function of $\mathbf{p}$. We put

$$c^+(\mathbf{p})\Phi = \bar{c}(\mathbf{p})\Phi \qquad (2.15)$$

where on the right hand side we have the same auxiliary function $\bar{c}(\mathbf{p})$, as in formula (2.13). We made it clear in our discussion of formulas (1.5)—(1.7) of § 1, that from an equality of type (2.15) in conjunction with commutation relation

$$[c(\mathbf{p}), c^+(\mathbf{p}')] = \delta^3(p - p') \qquad (2.16)$$

it follows that Φ ought to be a functional of the function $\bar{c}(\mathbf{p})$. In addition the operator $c(\mathrm{p})$ ought to be the operator of the functional derivative with respect to $\bar{c}(\mathrm{p})$. In other words

$$c(\mathrm{p})\,\Phi\{\bar{c}\} = \frac{\delta}{\delta\bar{c}(p)}\,\Phi\{\bar{c}\} \tag{2.17}$$

must hold. Formulas (2.15) and (2.17) replace in the general case formulas (2.4) and (2.5) of the elementary case of one-meson state.

In the representation under consideration the number operator has the form

$$\hat{n} = \int c^{+}(\mathrm{p})\,c(\mathrm{p})\,d^3p = \int \bar{c}(\mathrm{p})\,\frac{\delta}{\delta\bar{c}(\mathrm{p})}\,d^3p. \tag{2.18}$$

It is easy to verify that the eigenfunctions of $\hat{n}$ will be products of the functions $c(p)$: the functional

$$\mathrm{A}_n\{\bar{c}\} = A_n\bar{c}(\mathrm{p}_1)\,\bar{c}(\mathrm{p}_2)\ldots\bar{c}(\mathrm{p}_n) \tag{2.19}$$

(A_n being a normalizing factor) belongs to the eigenvalue n. In the particular case where the momenta of all the mesons are equal $\mathrm{p}_1 = \mathrm{p}_2\ldots = \mathrm{p}_n$, we obtain formula (2.7) of the elementary example. Of special significance is the functional of the vacuum A_0 (more precisely the functional of the free vacuum, since we are using operators of non-interacting fields). According to (2.19), the normalized vacuum functional $\mathrm{A}_0 = 1$ in the representation under consideration. Therefore (2.19) is equivalent to the expression

$$\mathrm{A}_n = A_n\,c^{+}(\mathrm{p}_1)\,c^{+}(\mathrm{p}_2)\ldots c^{+}(\mathrm{p}_n)\,\mathrm{A}_0. \tag{2.20}$$

The operators $c^{+}(\mathrm{p})$ and $c(\mathrm{p})$ are Hermitian conjugates by definition. Therefore, the scalar product $(\Omega_n, \Omega') \equiv \Omega_n^{*}\Omega'$ of the functionals Ω_n (formula (2.13) and of an arbitrary functional Ω' may be written in the form

$$(\Omega_n, \Omega') = \frac{1}{\sqrt{n!}}\left(\int \Psi_n(\mathrm{p}_1\ldots\mathrm{p}_n)\,c^{+}(\mathrm{p}_1)\ldots c^{+}(\mathrm{p}_n)\mathrm{A}_0\,d^3p_1\ldots d^3p_n\ \Omega'\right) =$$
$$= \frac{1}{\sqrt{n!}}\int \Psi_n(\mathrm{p}_1\ldots\mathrm{p}_n)\,(\mathrm{A}_0, c(\mathrm{p}_1)\ldots c(\mathrm{p}_n)\,\Omega')\,d^3p_1\ldots d^3p_n. \tag{2.21}$$

If we take into consideration that $c(\mathrm{p})\,\mathrm{A}_0 = 0$ and $\mathrm{A}_0^{*}c^{+}(\mathrm{p}) = 0$, then the matrix element in the right-hand side of (2.21) may also be represented in the form

$$(\mathrm{A}_0, c(\mathrm{p}_1)\ldots c(\mathrm{p}_n)\,\Omega') = \left.\frac{\delta^n\,\Omega'}{\delta\bar{c}(\mathrm{p}_1)\ldots\delta\bar{c}(\mathrm{p}_n)}\right|_{\bar{c}(\mathrm{p})=0}, \tag{2.22}$$

where, after performing the functional differentiation, we have to put $\bar{c}(p_i)=0$. In particular, for $\Omega'=\Omega'_m$, we find $(\Omega_n, \Omega'_m)=\delta_{nm}(\Omega_n, \Omega'_n)$ or in the general case

$$(\Omega, \Omega')=\sum_n \int \Psi_n(\mathbf{p}_1 \ldots \mathbf{p}_n)\Psi'_n(\mathbf{p}_1 \ldots \mathbf{p}_n)\, d^3p_1 \ldots d^3p_n. \tag{2.23}$$

If we put, $\Omega'=\Lambda_n$, in formula (2. 22) then we obtain for the normalizing constant A_n (formula (2.19)) the value $A_n=(n!)^{-1/2}$. Thus, series (2.13) represents the expansion of the generating functional in terms of the eigenfunctionals of the number of particles Λ_n. Since we assign to the coefficients $\Psi(\mathbf{p}_1 \ldots \mathbf{p}_n)$ of the expansion the meaning of probability amplitudes, $\Omega\{\overset{n}{\bar{c}}\}$ indeed coincides with the state vector Φ in a representation where $c^+(\mathbf{p})$ is the operator of multiplication by $\bar{c}(\mathbf{p})$ We may now define the amplitude $\Psi(\mathbf{p}_1 \ldots \mathbf{p}_n)$ as the quantity

$$\begin{aligned}\Psi_n(\mathbf{p}_1 \ldots \mathbf{p}_n) &= \frac{1}{\sqrt{n!}}(\mathrm{A}_0, c(\mathbf{p}_1)\ldots c(\mathbf{p}_n)\Omega)= \\ &= \frac{\delta^n \Omega\{\bar{c}\}}{\delta\bar{c}(\mathbf{p}_1)\ldots\delta\bar{c}(\mathbf{p}_n)} \frac{1}{\sqrt{n!}}\Bigg|_{\bar{c}=0}\end{aligned} \tag{2.24}$$

From (2.24) we obtain at once the relations between the functions Ψ_n and Ψ'_n of the functionals Ω and Ω', if $\Omega'=c^+(\mathbf{p})\Omega$.. Substituting Ω' in (2. 24) and carrying the function $\bar{c}(p)$, to the left side through functional differentiation, we have:

$$\Psi'_n(\mathbf{p}_1\mathbf{p}_2 \ldots \mathbf{p}_n)=\frac{1}{\sqrt{n}}[\Psi_{n-1}(\mathbf{p}_2\mathbf{p}_3 \ldots \mathbf{p}_n)\delta^3(p-p_1)]_p, \tag{2.25}$$

where $[\ ]_p$ denotes the symmetrization of the expression inside the brackets with respect to the variables $\mathbf{p}$ and $\mathbf{p}_j$:

$$[F(\mathbf{p}_1 \ldots \mathbf{p}_n\mathbf{p})]_p=F(\mathbf{p}_1 \ldots \mathbf{p}_n\mathbf{p})+F(\mathbf{p}_1 \ldots \mathbf{p}_{n-1}\mathbf{p}\mathbf{p}_n)+\ldots$$

If $\Omega''=c(\mathbf{p})\Omega$, , an additional derivative is added in (2.24), and we have:

$$\Psi''_n(\mathbf{p}_1 \ldots \mathbf{p}_n)=\sqrt{n+1}\,\Psi_{n+1}(\mathbf{p}\mathbf{p}_1 \ldots \mathbf{p}_n). \tag{2.26}$$

Let us find the eigenfunctionals of the annihilation operator $c(\mathbf{p})$. By (2.17) we have for the eigenvalue c'

$$c(\mathbf{p})F_{c'}\{\bar{c}\}=\frac{\delta}{\delta\bar{c}(\mathbf{p})}F_{c'}\{\bar{c}\}=c'(\mathbf{p})F_{c'}\{\bar{c}\},$$

or

$$F_{c'}\{\bar{c}\} \sim \exp\left[\int c'(\mathbf{p})\,\bar{c}(\mathbf{p})\,d^3p\right]. \qquad (2.27)$$

Let us now consider

$$(\Lambda_0, \exp\left[\int c(\mathbf{p})\,\bar{c}'(\mathbf{p})\,d^3p\right]\Omega). \qquad (2.28)$$

By expanding the exponent in a series and in view of definition (2.24) of the amplitude Ψ_n, we can convince ourselves that (2.28) is nothing else but the generating functional(of the function $\bar{c}'(\mathbf{p})$) for the amplitudes $\tilde{\Psi}_n = \sqrt{n!}\,\Psi_n$.

3. The method of functionals and Fermi statistics. An extension of the method of generating functional to the case of fields which conform to Fermi statistics meets with difficulties. The commutation relations (1.19)-(1.20) for the creation operators a^+, b^+ and annihilation operators a, b are of positive, instead of negative sign,as in the case of Bose-field operators, and the operators a^+ and b^+ anticommute:

$$\{a^+(\mathbf{p}), a^+(\mathbf{p}')\} = 0, \qquad \{a^+(\mathbf{p}), b^+(\mathbf{p}')\} = 0. \qquad (2.29)$$

In this case it is not possible to introduce a representation in which the effect of the operator $a^+(\mathbf{p})$ on the functional Ω would consist in multiplication by the function $\bar{a}(\mathbf{p})$: if we assume the relation

$$a^+(\mathbf{p})\,\Omega = \bar{a}(\mathbf{p})\,\Omega,$$

then the following equality must also hold

$$a^+(\mathbf{p})\,a^+(\mathbf{p}')\,\Omega = \bar{a}(\mathbf{p})\,\bar{a}(\mathbf{p}')\,\Omega = \bar{a}(\mathbf{p}')\,\bar{a}(\mathbf{p})\,\Omega,$$

in contradiction to commutation rule (2.29), from which moreover, it follows that $\bar{a}(\mathbf{p})^2 = 0$.

One possibility of generalization is connected with a formula analogous to formula (2.20). From (2.20), it follows that the expansion of Ω in a functional power series (1.15) is equivalent to an expansion in series in terms of products of the creation operators which act on the normalized vacuum functional Λ_0. The products

$$\left.\begin{array}{l} \Lambda_{nml} = (n!\,m!\,l!)^{-\frac{1}{2}}\, c^+(\mathbf{k}_1\ldots) c^+(\mathbf{k}_l)\, b^+(\mathbf{q}_m)\ldots b^+(\mathbf{q}_1)\, a^+(\mathbf{p}_n)\ldots a^+(\mathbf{p}_1)\Lambda_0, \\ {[a(\mathbf{p})\,\Lambda_0 = 0, \quad b(\mathbf{q})\,\Lambda_0 = 0, \quad c(\mathbf{k})\,\Lambda_0 = 0, \quad (\Lambda_0, \Lambda_0) = 1]} \end{array}\right\} \qquad (2.30)$$

are the eigenvectors of the number-of-particles operators $\int a^+(\mathbf{p})\,a(\mathbf{p})\,d^3p$ and $\int b^+(\mathbf{q})\,b(\mathbf{q})\,d^3q$ in the case of Fermi statistics. We denote by

$\Psi_{nml}(\mathbf{p}_1\ldots\mathbf{p}_n|\mathbf{q}_1\ldots\mathbf{q}_m|\mathbf{k}_1\ldots\mathbf{k}_l)$ the probability amplitude for the instant t, of finding in the field nucleons with momenta $\mathbf{p}_1\ldots\mathbf{p}_n$, m anti-nucleons with momenta $\mathbf{q}_1\ldots\mathbf{q}_m$ and l mesons with momenta $\mathbf{k}_1\ldots\mathbf{k}_l$. The function Ψ_{nml} is antisymmetric in the variables $\mathbf{p}_1\ldots\mathbf{p}_n;\ \mathbf{q}_1\ldots\mathbf{q}_m$ and symmetric in the variables $\mathbf{k}_1\ldots\mathbf{k}_l$. The vector of state Ω will then be equal to

$$\Omega=\sum_{nml}^{\infty}\Omega_{nml}=\sum_{nml}(n!\,m!\,l!)^{-1/2}\int\Psi_{nml}(\mathbf{p}_1\ldots|\mathbf{q}_1\ldots|\mathbf{k}_1\ldots)\times$$
$$\times c^+(\mathbf{k}_1)\ldots c^+(\mathbf{k}_l)\,b^+(\mathbf{q}_m)\ldots b^+(\mathbf{q}_1)\,a^+(\mathbf{p}_n)\ldots a^+(\mathbf{p}_1)\,\Lambda_0. \tag{2.31}$$

For the quantity Ψ_{nml} we obtain:

$$\Psi_{nml}(\mathbf{p}_1\ldots\mathbf{p}_n|\mathbf{q}_1\ldots\mathbf{q}_m|\mathbf{k}_1\ldots\mathbf{k}_l)=$$
$$=(n!\ m!\ l!)^{-1/2}(\Lambda_0,\ a^+(\mathbf{p}_1)\ldots a^+(\mathbf{p}_n)\,b^+(\mathbf{q}_1)\ldots$$
$$\ldots b^+(\mathbf{q}_m)\,c^+(\mathbf{k}_1)\ldots c^+(\mathbf{k}_l)\,\Omega). \tag{2.32}$$

From (1.4) we can readily find the result of the action of the creation a^+ and annihilation a operators on Ω. If $\Omega'=a^+(\mathbf{p})\,\Omega$, the amplitudes Ψ'_{nml} of Ω' are connected with the amplitudes Ψ_{nml} of Ω by a relation which is obtained by substituting Ω for Ω* in (2.32) and by carrying the operator $a^+(\mathbf{p})$ to the left, to Λ_0. In view of the fact that $(\Lambda_0,\ a^+(\mathbf{p})\,\Omega)=0$ for any Ω, we obtain:

$$\Psi'_{nml}(\mathbf{p}_1\ldots\mathbf{p}_n|\ldots|\ldots)=\frac{1}{\sqrt{n}}\{\delta^3(p-p_1)\,\Psi_{n-1\,ml}(\mathbf{p}_2\ldots\mathbf{p}_n|\ldots|\ldots)-$$
$$-\delta^3(p-p_2)\,\Psi_{n-1\,ml}(\mathbf{p}_1,\ \mathbf{p}_3\ldots\mathbf{p}_n|\ldots|\ldots)+\ldots$$
$$\ldots-(-1)^n\,\delta^3(p-p_n)\,\Psi_{n-1\,ml}(p_1\ldots p_{n-1}|\ldots|\ldots)\}. \tag{2.33}$$

If $\Omega''=a(\mathbf{p})\,\Omega$, the corresponding amplitude Ψ''_{nml} is determined directly from (2.32)

$$\Psi''_{nml}(\mathbf{p}_1\ldots\mathbf{p}_n|\ldots|\ldots)=\sqrt{n+1}\,\Psi_{n+1\,ml}(\mathbf{p}\,\mathbf{p}_1\ldots\mathbf{p}_n|\ldots|\ldots). \tag{2.34}$$

However, the representation of the vector of state in the form (2.31), constitutes, in fact, a departure from the of functionals and a return to the operator method of second quantization. In order to develop the method of functionals for Fermi statistics in accordance with the commutation relations (2.29), it is necessary to define a system of such quantities $\bar{a}(\mathbf{p})$ and $\bar{b}(\mathbf{q})$, depending on the momentum $\mathbf{p}$ of the nucleon, and $\mathbf{q}$, of the antinucleon, which anticommute with each other

$$\left.\begin{aligned}&\{\bar{a}(\mathbf{p}),\ \bar{a}(\mathbf{p}')\}=0,\quad \{\bar{b}(\mathbf{q}),\ \bar{b}(\mathbf{q}')\}=0,\\ &\{\bar{a}(\mathbf{p}),\ \bar{b}(\mathbf{q})\}=0\end{aligned}\right\} \tag{2.35}$$

----

* [Transl. note: It should be Ω ', not Ω as in text].

and with the operators of the nucleon field, and which are also proportional to the unitary matrix of the space of nucleon, antinucleon and meson occupation numbers. This can be done by means of the external-sources representation $\eta(x)$ and $\bar{\eta}(x)$, of the nucleon field, introduced by Schwinger /9/. The sources $\eta(x)$ and $\bar{\eta}(x)$ may be interpreted as quantities referring to some other given field. Since the operators of different spinor fields anticommute with each other, the sources η and $\bar{\eta}$ anticommute with the operators Ψ and $\bar{\Psi}$. In addition, $\{\eta, \bar{\eta}\}=0$, since the spinor field $\eta, \bar{\eta}$ is assumed as given. The quantities $\bar{a}(k)$ and $\bar{b}(q)$, for which (2.35) holds can be identified with the components of the expansion of the negative-frequency parts of the external sources in the three-dimensional Fourier integral:

$$\left.\begin{aligned} \bar{\eta}^{(-)}(x) &= \frac{1}{(2\pi)^{3/2}} \int \bar{a}(\mathbf{p})\, \bar{U}(\mathbf{p})\, e^{-ipx} d^3p, \\ \eta^{(-)}(x) &= \frac{1}{(2\pi)^{3/2}} \int \bar{b}(\mathbf{p})\, \bar{V}(\mathbf{p})\, e^{-ipx} d^3p, \end{aligned}\right\} \tag{2.36}$$

where $V(\mathbf{p})$ and $U(\mathbf{p})$ are the Dirac spinors for the given external field, analogous to the spinors $v(\mathbf{p})$ and $u(\mathbf{p})$ in formulas (1.17)—(1.18). We may now pass to a representation where $a^+(\mathbf{p})$ and $b^+(\mathbf{q})$, are the multiplication operators:

$$a^+(\mathbf{p})\,\Omega = \bar{a}(\mathbf{p})\,\Omega; \quad b^+(\mathbf{p})\,\Omega = \bar{b}(\mathbf{p})\,\Omega. \tag{2.37}$$

Since $\bar{a}$ and $\bar{b}$ anticommute, the order of the factors is essential in defining the functional derivatives with respect to $\bar{a}$ and $\bar{b}$. By definition,

$$\left.\begin{aligned} \delta\Omega\{\bar{a}\} &= \int \delta\bar{a}(\mathbf{p}) \frac{\delta\Omega}{\delta\bar{a}(\mathbf{p})} d^3p, \\ \delta\Omega\{\bar{b}\} &= \int \frac{\delta\Omega}{\delta\bar{b}(\mathbf{p})} \delta\bar{b}(\mathbf{p})\, d^3p. \end{aligned}\right\} \tag{2.38}$$

The operators of the functional derivatives with respect to $\bar{a}$ and $\bar{b}$ anticommute:

$$\left\{\frac{\delta}{\delta\bar{a}(\mathbf{p})}, \frac{\delta}{\delta\bar{b}(\mathbf{q})}\right\} = 0, \quad \left\{\frac{\delta}{\delta\bar{a}(\mathbf{p})}, \frac{\delta}{\delta\bar{a}(\mathbf{p}')}\right\} = 0.$$

It follows from (2.38) that

$$\left\{\frac{\delta}{\delta\bar{a}(\mathbf{p})}, \bar{a}(\mathbf{p}')\right\} = \delta^3(p-p'), \quad \left\{\frac{\delta}{\delta\bar{b}(\mathbf{p})}, \bar{b}(\mathbf{p}')\right\} = \delta^3(p-p').$$

Hence it follows that, in complete accordance with the method of functionals for Bose statistics, we may put

$$a(\mathbf{p}) = \frac{\delta}{\delta\bar{a}(\mathbf{p})}\,\Omega; \quad b(\mathbf{p}) = \frac{\delta}{\delta\bar{b}(\mathbf{p})}\,\Omega. \tag{2.39}$$

By analogy with Bose statistics, we may conclude that in the general case of interacting nucleon and meson fields, the expansion of the vector of state as a functional of $\bar{a}$, $\bar{b}$ and $\bar{c}$, differs from (2.31) in that the products $c^+(\mathbf{k}_1)\ldots c^+(\mathbf{k}_e)\, b^+(\mathbf{q}_m)\ldots b^+(\mathbf{q}_1)\, a^+(\mathbf{p}_n)\ldots a^+(\mathbf{p}_1)\,\Lambda_0$ are replaced by $c(\bar{\mathbf{k}}_1)\ldots \bar{c}(\mathbf{k}_e)\,\bar{b}(\mathbf{q}_m)\ldots$ $\ldots\bar{b}(\mathbf{q}_1)\,\bar{a}(\mathbf{p}_n)\ldots a(\mathbf{p}_1)$.

Let us consider the operator

$$R\{\bar{a}',\ \bar{b}'\} = \exp\{\int[\bar{a}'(\mathbf{p})\,a(\mathbf{p}) + b(\mathbf{p})\,\bar{b}'(\mathbf{p})]\,d^3p\}, \tag{2.40}$$

which is a functional of $\bar{a}'(\mathbf{p})$ and $\bar{b}'(\mathbf{p})$. We have:

$$\frac{\delta}{\delta\bar{a}'(\mathbf{p})}R = Ra(\mathbf{p}), \qquad \frac{\delta^n R}{\delta\bar{a}'(\mathbf{p}_1)\ldots\delta\bar{a}'(\mathbf{p}_n)} = Ra(\mathbf{p}_n)\ldots a(\mathbf{p}_1)$$

and analogous formulas for the functional derivatives with respect to $\bar{b}(\mathbf{q})$. By comparing these formulas with expression (2.32) for the probability amplitudes, we can conclude that the generating functional $\tilde{\Omega}$ for the amplitudes $\tilde{\Psi}_{nml} = \sqrt{n!m!l!}\,\Psi_{nml}$ can be represented in the form

$$\tilde{\Omega}\{\bar{a}',\bar{b}',\bar{c}'\} = (\Lambda_0,\ Re^{\int\bar{c}'(\mathbf{k})c(\mathbf{k})\,d^3k}\,\Omega). \tag{2.41}$$

Instead of formulas (2.32) for the probability amplitudes Ψ_{nml} we may then write:

$$\Psi_{nml}(\mathbf{p}_1\ldots\mathbf{p}_n\,|\,\mathbf{q}_1\ldots\mathbf{q}_m\,|\,\mathbf{k}_1\ldots\mathbf{k}_l) = \\ = \frac{1}{\sqrt{n!\,m!\,l!}}\,\frac{\delta^{n+m+l}\Omega}{\delta\bar{a}(\mathbf{p}_1)\ldots\delta\bar{a}(\mathbf{p}_n)\,\delta\bar{b}(\mathbf{q}_1)\ldots\delta\bar{b}(\mathbf{q}_m)\,\delta\bar{c}(\mathbf{k}_1)\ldots\delta\bar{c}(\mathbf{k}_l)}\Big|_{\bar{a}=\bar{b}=\bar{c}=0}. \tag{2.42}$$

Just as in the derivation of formulas (2.23), we can show that the condition that the operators a and a^+ , b and b^+ are Hermitian conjugates leads to the definition of the scalar product of the two operators Ω and Ω':

$$(\Omega',\Omega) = \sum_{nml}\int \psi^{*\prime}_{nml}(\mathbf{p}_1\ldots\mathbf{p}_n\,|\,\mathbf{q}_1\ldots\mathbf{q}_m\,|\,\mathbf{k}_1\ldots\mathbf{k}_l)\,\psi_{nml}(\mathbf{p}_1\ldots\mathbf{p}_n\,|\,\mathbf{q}_1\ldots\mathbf{q}_m\,|\,\mathbf{k}_1\ldots\mathbf{k}_l)\times \\ \times d^3p_1\ldots d^3p_n d^3q_1\ldots d^3q_m d^3k_1\ldots d^3k_l. \tag{2.43}$$

4. The equations for the functional of state. It is necessary for deducing the equation which is satisfied by the functional of state which we considered in § 2, to express the energy operator H in terms of the creation operators $a^+(\mathbf{p}),\ b^+(\mathbf{q}),\ c^+(\mathbf{k})$ and the annihilation operators $a(\mathbf{p}),\ b(\mathbf{q}),\ c(\mathbf{k})$ of various particles:

$$H = H(a,\ b,\ c,\ a^+,\ b^+,\ c^+). \tag{2.44}$$

As was made clear in § 2, it is possible to introduce, in the method of functionals, such a representation, in which the creation operators are operators of multiplication by the auxiliary quantities $\bar{a}(\mathbf{p}), \bar{b}(\mathbf{q}), \bar{c}(\mathbf{k})$. Then the effect of the annihilation operators in this representation consists in forming the functional derivative of the state vector Ω with respect to these auxiliary quantities. In order to obtain the equation of motion for the functional Ω, it is therefore necessary to make the following substitution in expression (2.44):

$$\left.\begin{array}{l} a^+(\mathbf{p}) \to \bar{a}(\mathbf{p}),\ b^+(\mathbf{q}) \to \bar{b}(\mathbf{q}),\ c^+(\mathbf{k}) \to \bar{c}(\mathbf{k}) \\ a(\mathbf{p}) \to \delta/\delta\bar{a}(\mathbf{p}),\ b(\mathbf{q}) \to \delta/\delta\bar{b}(\mathbf{q}),\ c(\mathbf{k}) \to \delta/\delta\bar{c}(\mathbf{k}), \end{array}\right\} \tag{2.45}$$

after which the Schroedinger equation can be written in the form

$$H(\bar{a},\ \bar{b},\ \bar{c},\ \delta/\delta\bar{a},\ \delta/\delta\bar{b},\ \delta/\delta\bar{c}) = i\partial\Omega/\partial t. \tag{2.46}$$

From functional equation (2.46), the equation for the probability amplitudes Ψ_{nml} (2.32) can be obtained.

For the interaction of a neutral pseudoscalar meson field with a nucleon field, the energy operator H has, in the case of pseudoscalar coupling, the following form:

$$\left.\begin{array}{l} H = H_1 + H_2 + H_{12}, \\ H_1 = \frac{1}{2}\int d^3x\,(\pi^2 + (\nabla\varphi)^2 + \mu^2\varphi^2), \\ H_2 = \int d^3x\,(\bar{\psi}(\gamma\nabla)\psi + m\bar{\psi}\psi), \\ H_{12} = ig\int d^3x\bar{\psi}\gamma_5\psi\varphi. \end{array}\right\} \tag{2.47}$$

By using the Fourier expansion of the field operators (1.16-1.18), we obtain in view of (2.45), the following expressions for H_1 and H_2 :

$$H_1 = \int d^3k k_0 \bar{c}(\mathbf{k})\,\delta/\delta\bar{c}(\mathbf{k}),$$

$$H_2 = \sum_{i=1}^{2}\int d^3p E(p)\,\{\bar{a}_i(\mathbf{p})\,\delta/\delta\bar{a}_i(\mathbf{p}) + \bar{b}_i(\mathbf{p})\,\delta/\delta\bar{b}_i(\mathbf{p})\},$$

$$k_0 = +\sqrt{\mu^2 + \mathbf{k}^2},\quad p_0 = E(p) = \sqrt{m^2 + \mathbf{p}^2}.$$

For H_{12} we obtain quite a cumbersome expression, consisting of eight terms which correspond to the different combinations of the creation and annihilation operators of the meson and nucleon fields. The equation for the probability amplitudes (2.31) can be obtained by using expansion (2.32) for the functional of state Ω and by comparing, in equation (2.46), the terms which contain the same number of auxiliary quantities $\bar{a}$, $\bar{b}$, $\bar{c}$. We shall illustrate these statements by a simpler example, when the interaction operator of the meson and nucleon fields has the form:

$$H_{12} = ig\gamma_5\varphi(x).$$

In this case we do not take into account the closed loops which correspond to the creation of virtual pairs of nucleons-antinucleons.

The equation for the wave functional assumes the simpler form:

$$(\gamma_\mu \partial/\partial x_\mu + m)\,\Omega(x) = -ig\gamma_5\varphi(x)\,\Omega(x), \tag{2.48}$$

where the operator $\varphi(x)$ can be written in the form

$$\varphi(x) = \frac{1}{(2\pi)^{3/2}} \int \frac{d^3k}{\sqrt{2k_0}} \left(e^{ik_\mu x_\mu} \delta/\delta\bar{c}(\mathbf{k}) + e^{-ik_\mu x_\mu} \bar{c}(\mathbf{k}) \right). \tag{2.49}$$

From (2.48) and (2.49) follows

$$(\gamma_\mu \partial/\partial x_\mu + m)\,\Omega = \int d^3k \,\{G^+(k)\,\delta\Omega/\delta\bar{c}(\mathbf{k}) - G(k)\,\bar{c}(k)\,\Omega\}, \tag{2.50}$$

where

$$G(k) = \frac{i}{(2\pi)^{3/2}}\, g\, \frac{e^{-ik_\mu x_\mu}}{\sqrt{2k_0}} \gamma_5. \tag{2.51}$$

In equation (2.50) the coefficients $G(k)$ and $G^+(k)$ are explicitly time-dependent. This dependence can be eliminated by passing to the Schroedinger representation

$$L_0 = \exp(-iH_1 t)\, L \exp(iH_1 t), \quad \Omega = \exp(+iH_1 t)\,\Omega_0. \tag{2.52}$$

In that case the operators $\delta/\delta\bar{c}(\mathbf{k})$, $\bar{c}(\mathbf{k})$ are transformed according to

$$\left.\begin{aligned} \exp(-iH_1 t)\,\delta/\delta\bar{c}(k)\exp(iH_1 t) &= \exp(ik_0 t)\,\delta/\delta\bar{c}(k), \\ \exp(-iH_1 t)\,\bar{c}(k)\exp(iH_1 t) &= \exp(-ik_0 t)\,\bar{c}(k). \end{aligned}\right\} \tag{2.53}$$

As a result we obtain for the functional Ω the following equation*:

$$\begin{aligned} &(\gamma_\mu \partial/\partial x_\mu + m + \gamma_4 \textstyle\int k_0 \bar{c}(k)\,\delta/\delta\bar{c}(k))\,\Omega = \\ &\qquad = \int d^3k \,\{G_0^+(k)\,\delta\Omega/\delta\bar{c}(\mathbf{k}) - G_0(k)\,\bar{c}(\mathbf{k})\,\Omega\}. \end{aligned} \tag{2.54}$$

----

* We denote the functional of state, as hitherto, by Ω instead of Ω_0.

From equation (2.54) we obtain, in view of (2.13) and (2.14), the following equation for the probability amplitude:

$$\left(\gamma_\mu \partial/\partial x_\mu + m + \gamma_4 \sum_k N_k k_0\right) \psi_n (x, k_1 \ldots k_n) =$$
$$= \sqrt{n+1} \int d^3k G_0^+ (k) \psi_{n+1} (x, k, k_1 \ldots k_n) -$$
$$- \frac{1}{\sqrt{n}} \left\{G_0 (k) \psi_{n-1} (x, k_2 \ldots k_n)\right\}_{\mathrm{sym}}. \tag{2.55}$$

Equation (2.55) relates the amplitude ψ_n with the amplitudes ψ_{n+1} and ψ_{n-1}, for which analogous equations can be written. As a result we obtain an infinite system of "linked" equations. For the one-meson approximation, e.g., we obtain:

$$\left.\begin{aligned} &(\gamma_\mu \partial/\partial x_\mu + m) \psi_0 (x) = \int d^3k G_0^+ (k) \psi_1 (k, x), \\ &(\gamma_\mu \partial/\partial x_\mu + m + \gamma_4 k_0) \psi_1 = - G_0 (k) \psi_0 (x). \end{aligned}\right\} \tag{2.56}$$

Equations (2.55) can also be directly deduced from expression (2.32) for the probability amplitudes, if we were to use relation*

$$i\partial\psi_n/\partial t = (\Lambda_0, c^+(k_1) \ldots c^+(k_n) H_{int}\Omega). \tag{2.57}$$

The right-hand side of (2.57) can be readily computed on the basis of (2.45) and (2.47). We can obtain, in an analogous way, the equations for probability amplitudes containing operators of the Fermi field.

The system of linked equations (2.55) was obtained by V.A. Fock in 1934. From these equations in approximation (2.56), formulas of Breit and Moller were obtained and the natural breadth of spectral lines studied. These equations were subsequently used by A.A. Smirnov /5/ and A.G. Vlasov /6/ in the study of the interaction between the electron and the electromagnetic field. The theory of the natural breadth of spectral lines in a two-photon approximation was considered in reference /7/. The method of linked equations has acquired a special practical importance in meson theory, since in the latter the perturbation theory is useless. Equations (2.55) were obtained in meson theory, irrespective of earlier work in quantum electrodynamics, by I.E. Tamm /14/ and S.M. Dancoff /15/, and used by them for the investigation of nuclear forces. The system of equations (2.55) can be solved by the perturbation method and by the method of cut-off equations. The latter is known in literature under the name of the Tamm-Dancoff method and will be dealt with in the following.

----

* We are using the interaction representation.

It is assumed that in solving systems of equations of type (2.55) one can neglect all the amplitudes which describe the states of the field which have N + 1, N + 2 mesons and so on. In other words, in this approximation one retains in expansion $\Omega = \sum_{n=0}^{\infty} \Omega_n$, N + 1 subterms and the rest are assumed equal to zero. As a result we obtain a "cut-off" system of equations which must then be solved exactly (e.g., by numerical integration) with assigned boundary conditions. In the one-meson approximation (2.56) it is possible to eliminate the amplitude ψ_1 from the second equation and substitute it in the first; as a result we obtain one integral equation. In higher approximations, such a procedure may be followed only under the assumption that the probability amplitude which corresponds to the smallest possible number of particles in the given problem, is the fundamental one and the other amplitudes play the role of corrections*. We can then obtain one integral equation for the fundamental amplitude. Such a method, however, constitutes an oversimplification of the problem since, in that case, the kernel will be represented by a series in terms of the constant g^2, the different terms of the series being comparable in magnitude, and therefore such an expansion is useless for an investigation, even in an asymptotic sence.

When solving the system of equations (2.55) by the perturbation method /25/, it should be noted that each of these equations is a non-homogeneous Dirac equation of the type

$$D(x)\psi = i(\gamma_\mu \partial/\partial x_\mu + m)\psi(x) = L(x). \tag{2.58}$$

The general solution of this equation can be represented in the form

$$\left.\begin{aligned} \psi(x) = \varphi(x) + \int K(x, x')\, L(x')\, d^4x', \\ D(x)\varphi(x) = 0\,. \end{aligned}\right\} \tag{2.59}$$

Green's function K(x, x') ought to be chosen in accordance with the positron theory. The interaction described by the term L(x) should be adiabatically introduced at $t = -\infty$ and removed at $t = +\infty$, attaining its full magnitude throughout the entire finite-time interval. By successive elimination of amplitudes, it is possible to arrive at a single integral equation which corresponds to the integral equation of the Feynman theory. In such a method of solution, the renormalization is performed in the same way as in the perturbation theory.

Along with the probability amplitudes which are defined in the momentum space, we can also consider probability amplitudes in the coordinate

----

* The Levy-Klein method /17, 18/, see also /7/.

space /13, 20, 24/. The latter can be used to investigate the possibility of renormalization within the framework of the method of cut-off equations /4, 16/ and for the comparison of probability amplitudes with relativistic wave functions, which will be considered in § 4. For the introduction of the probability amplitudes in coordinate space, we represent the operators of the non-interacting fields $\varphi(x)$ and $\psi(x)$ in the form of a sum of terms, containing only positive and only negative frequencies:

$$\varphi(x)=\varphi^{(+)}(x)+\varphi^{(-)}(x),\quad \psi(x)=\psi^{(+)}(x)+\psi^{(-)}(x), \tag{2.60}$$

where

$$\varphi^{(\pm)}(x)=\int d\sigma_\mu(x')\left\{\Delta^{(\pm)}(x-x')\overleftrightarrow{\partial/\partial x'_\mu}\varphi(x')\right\}, \tag{2.61a}$$

$$\psi^{(\pm)}(x)=\int d\sigma_\mu(x')\,S^{(\pm)}(x-x')\,\gamma_\mu\psi(x'). \tag{2.61b}$$

The operators $\varphi^{(\pm)}(x)$, $\psi^{(\pm)}(x)$ satisfy the following commutation relations:

$$[\varphi^{(+)}(x),\ \varphi^{(-)}(x')]=i\,\Delta^{+}(x-x')=-\,i\Delta^{(-)}(x-x'). \tag{2.62}$$

$$\left\{\psi_\lambda^{(+)}(x),\ \overline{\psi_\mu^{(+)}}(x')\right\}=-\,iS_{\lambda\mu}^{(+)}(x-x'), \tag{2.63}$$

$$\Delta^{\pm}(x)=\mp\, i/(2\pi)^3\int\frac{d^3k}{2k_0}\exp(\pm\, ix_\mu x_\mu), \tag{2.64}$$

$$S^{(\pm)}(x)=(\gamma_\mu\,\partial/\partial x_\mu-m)\Delta^{(\pm)}(x). \tag{2.65}$$

The annihilation operators $a(\mathbf{p})$, $b(\mathbf{q})$ and $c(\mathbf{k})$ are expressed in terms of the field operators $\psi(x)$, $\overline{\psi}(x)$, $\varphi(x)$ in accordance with formulas (1.12) and (1.15). In these expressions only the positive-frequency terms are contributing to the corresponding integrals, and the required creation operators $a^+(\mathbf{p})$, $b^+(\mathbf{q})$, $c^+(\mathbf{k})$ can be therefore represented by the following formulas*:

$$\left.\begin{aligned}a^+(\mathbf{p})&=\int d\sigma_\mu\,\overline{\psi}^{(-)}(x)\,\gamma_\mu\,u^{(p)}(x),\\ b^+(\mathbf{q})&=\int d\sigma_\mu\,\overline{v}^{(q)}(x)\,\gamma_\mu\,\psi^{(-)}(x),\end{aligned}\right\} \tag{2.66}$$

$$c^+(\mathbf{k})=-\,i\int d\sigma_\mu\,\varphi^{(-)}(x)\frac{\overleftrightarrow{\partial}}{\partial x_\mu}\,f^{(k)}(x). \tag{2.67}$$

By means of (2.67) we now transform expression (2.31) for the functional Ω to the coordinate space. For the sake of brevity this transformation is carried out only for the amplitudes of a neutral Bose-field**.

---

* See formulas (1.16)—(1.18)

** Analogous transformations for a charged Bose-field and for Fermi-fields can be found in reference /24/.

The functional $\Omega_1 = \int d^3k\, \psi(k)\, c^+(k)\, \Lambda_0$ transforms, by virtue of (2.67), to the following form:

$$\Omega_1 = -i \int\int d^3k\, d\sigma_\mu\, \Psi_1(k)\, \varphi^{(-)}(x) \frac{\overleftrightarrow{\partial}}{\partial x_\mu} f^{(k)}(x) = $$
$$= -i \int d\sigma_\mu\, (\varphi)^-(x) \frac{\overleftrightarrow{\partial}}{\partial x_\mu} \Psi_1(x)\, \Lambda_0, \tag{2.68}$$

where $\psi_1(x) = \int \psi_1(k)\, f^{(k)}(x)\, d^3k$.

By analogy

$$\left.\begin{aligned} \Omega_n = \frac{(-1)^n}{\sqrt{n!}} \int d\sigma_{\mu_1}(x_1) \ldots d\sigma_{\mu_n}(x_n) \prod_j \varphi^{(-)}(x_j) \frac{\overleftrightarrow{\partial}}{\partial x_{\mu_j}} \times \\ \times \psi_n(x_1 \ldots x_n, \sigma)\, \Lambda_0, \\ \psi_n(x_1 \ldots x_n, \sigma) = \int \psi(k_1 \ldots k_n)\, f^{(k_1)}(x_1) \ldots f^{(k_n)}(x_n)\, d^3x_1 \ldots d^3x_n. \end{aligned}\right\} \tag{2.69}$$

The amplitude $\psi_n(x_1 \ldots x_n)$ is symmetric in its arguments and satisfies the following relation:

$$\psi_n(x_1 \ldots x_n \sigma) = \int d\sigma_\mu(x'_j) \left(\Delta^{(+)}(x_j - x'_j) \frac{\overleftrightarrow{\partial}}{\partial x'_j} \right) \times$$
$$\times \psi_n(x_1 \ldots x'_j \ldots x_n, \sigma), \tag{2.70}$$

which follows from the fact that expression (2.70) is independent of the choice of the surface $\sigma(x)$ and from the circumstance that the function $\Delta^{(+)}(x - x')$ is a positive-frequency solution of the Klein-Fock equation.

The operators $\varphi^{(+)}(x)$ and $\varphi^{(-)}(x)$ have the following representation: If we denote, as in § 2, $\Omega'' = \varphi^{(+)}(x)\Omega$, then the functions

$$\psi''_n(x_1 \ldots x_n, \sigma) \text{ и } \psi_n(x, x_1 \ldots x_n, \sigma)$$

will be related in the following way:

$$\psi''_n(x_1 \ldots x_n, \sigma) = (n+1)^{1/2} \int d\sigma_\mu(x')\, \Delta^{(+)}(x - x') \frac{\overleftrightarrow{\partial}}{\partial x'_\mu} \Psi_n \times$$
$$\times (x', x_1 \ldots x_n, \sigma) = (n+1)^{1/2}\, \psi_{n+1}(x, x_1 \ldots x_n, \sigma). \tag{2.71}$$

By analogy, proceeding from the relation $\varphi^{(-)}(x)\Omega = \Omega'$, we obtain:

$$\Psi'_n(x_1 \ldots x_n) = \frac{i}{\sqrt{n!}} \left\{ \Delta^{(+)}(x - x_1)\, \psi_{n-1}(x_2 \ldots x_n) \right\}_{\text{sym}} \tag{2.72}$$

The basic vectors in coordinate space will be given by the following expressions:

$$\Lambda_n = \frac{1}{\sqrt{n!}} \varphi^{(-)}(x_1) \ldots \varphi^{(-)}(x_n) \Lambda_0. \qquad (2.73)$$

Hence the number operator can be put in the form:

$$N = i^{-1} \int d\sigma_\mu \left(\varphi^{(-)}(x) \frac{\overleftrightarrow{\partial}}{\partial x_\mu} \varphi^{(+)}(x) \right). \qquad (2.74)$$

It can be readily seen that $N\Lambda_n = n\Lambda_n$. Thus the expansion of the functional Ω in terms of the functionals Ω_n is in fact an expansion in terms of the eigenfunctions of the operator of the number of particles. The "coefficients" of this expansion are the probability amplitudes of finding n particles in coordinate space. The functionals Ω_n and $\Omega_{n'}$ with different indexes are mutually orthogonal, in which case, as it can be readily seen, the following relation holds:

$$(\Omega, \Omega) = (\Lambda_0, \Lambda_0) + \sum_{n=1}^{\infty} (-1)^n \int d\sigma_{\mu_1}(x_1) \ldots d\sigma_{\mu_n}(x_n) \times$$
$$\times \psi_n^*(x_1 \ldots x_n, \sigma) \prod_j \frac{\overleftrightarrow{\partial}}{(\partial x_j)_{\mu_j}} \psi_n(x_1 \ldots x_n, \sigma). \qquad (2.75)$$

By means of formulas (2.71) and (2.72) it can be shown that

$$(\Omega_n, \varphi^{(+)}(x)\, \Omega_{n+1}) = (\varphi^{(-)}(x)\, \Omega_n, \Omega_{n+1}),$$

i.e., that $\varphi^{(+)}(x)$ and $\varphi^{(-)}(x)$ are hermitian conjugates.

From expression (2.72) follows that

$$\psi_n(x_1 \ldots x_n) = \frac{1}{\sqrt{n!}} (\Lambda_0, \varphi^{(+)}(x_1) \ldots \varphi^{(+)}(x_n)\, \Omega\{\sigma\}). \qquad (2.76)$$

In the general case, the expression for the probability amplitude in coordinate space, describing a system of l nucleons, m antinucleons and n mesons can be written (omitting a numerical coefficient) in the form

$$\psi_{lmn} = (x_1 \ldots x_l \,|\, y_1 \ldots y_m \,|\, z_1 \ldots z_n) =$$
$$= (\Lambda_0, \bar{\psi}^{(+)}(x_1) \ldots \bar{\psi}^{(+)}(x_l);\ \psi^{(+)}(y_1) \ldots \psi^{(+)}(y_m);\ \varphi^{(+)}(z_1) \ldots \varphi^{(+)}(z_n)\, \Omega\{\sigma\}). \qquad (2.77)$$

In the following we shall assume that in expression (2.77) the vector of state Ω is taken in the interaction representation:

$$\frac{i\delta\Omega\{\sigma\}}{\delta\sigma(x)} = H_{int}(x)\,\Omega\{\sigma\}. \tag{2.78}$$

We can then obtain the equation for amplitudes in coordinate representation in the same way as the equation for amplitudes (2.32):

$$\begin{aligned}&\frac{i\delta}{\delta\sigma(x)}\psi(x_1\ldots x_l\,|\,y_1\ldots y_m\,|\,z_1\ldots z_n) =\\ &= (\Lambda_0,\ \overline{\psi}^{(+)}(x_1)\ \ldots\ \psi^{(+)}(y_1)\ \ldots\ \varphi^{(+)}(z_1)\ \ldots\ H_{int}\Omega).\end{aligned} \tag{2.79}$$

In that case the right-hand side of the equation should be represented, by virtue of (2.63), in the form of N products, and thus the amplitude will be related with the other amplitudes, for which analogous equations can be written; we obtain, e.g., for a one-meson amplitude, and taking into account that $H_{int} = ig\gamma_5\varphi$:

$$\begin{aligned}&\frac{i\delta}{\delta\sigma(x)}(\Lambda_0,\ \varphi^{(+)}(x_1)\,\Omega) = -\,g\,(\gamma_5)_{\alpha\beta}\,\{\Delta^{+}(x_1 - x)\times\\ &\times(\Lambda_0,\ \overline{\psi}_\alpha^{(+)}\,\psi_\beta^{(+)}(x)\,\Omega) - i\,(\Lambda_0,\ \overline{\psi}_\alpha^{(+)}(x)\,\psi_\beta^{(+)}(x)\,\varphi^{(+)}(x_1)\,\varphi^{(+)}(x)\,\Omega).\end{aligned} \tag{2.80}$$

For a one-nucleon amplitude we have

$$\begin{aligned}&\frac{i\delta}{\delta\sigma(x)}(\Lambda_0,\ \psi_\lambda^{(+)}(x_1)\,\Omega) = g\,(\gamma_5)_{\alpha\beta}\,\{S_{\lambda\alpha}^{(+)}(x_1 - x)\times\\ &\times(\Lambda_0,\ \psi_\beta^{(+)}(x)\,\varphi^{(+)}(x)\,\Omega) - i\,(\Lambda_0,\ \overline{\psi}_\alpha^{(+)}(x)\,\psi_\lambda^{(+)}(x_1)\,\psi_\beta^{(+)}(x)\,\varphi^{(+)}(x)\,\Omega).\end{aligned} \tag{2.81}$$

The amplitudes (2.77) were considered by Cini /16/ in an investigation of a renormalization possibility in the cut-off method. A critique of Cini's method and a detailed study of equations (2. 55) can be found in reference /4/.

Let us pass to the problem of the generating functional for the amplitudes

$$\tilde{\psi}_{lmn} = \sqrt{l!m!n!}\,\psi_{lmn}.$$

The latter can be written in the form (see also (2.41))

$$\begin{aligned}&F = (\Lambda_0,\ \exp\left\{\int d^3x\,(\eta(x)\,\psi^{(+)}(x) + \psi^{(-)}(x)\,\eta\ (x) + c\,(x)\,\varphi^{(+)}(x)\right\}\Omega) =\\ &= \sum_{l,\,m,\,n}\frac{1}{m!n!l!}F_{mnl}\,\Lambda_0 \equiv (\Lambda_0,\ M\Omega),\end{aligned} \tag{2.82}$$

where

$$\begin{aligned}F_{lmn}\{\eta,\ \eta^*,\ c|\,\} = \int\psi(x_1\ldots x_l\,|\,y_1\ldots y_m\,|\,z_1\ \ldots\ z_n)\times\\ \times\,\eta^*(x_1)\ldots\eta(y)\ldots c(z)\ldots d^3x\ldots d^3y\ldots d^3z\ldots\end{aligned} \tag{2.83}$$

F_{lmn} is a functional with respect to the functions $c(z)$ and to the quantities $\eta^{*}(x)$ and $\eta(y)$ which anticommute with each other. The equation of motion for the generating functional can be written in the form

$$\frac{i\delta F}{\delta\sigma(x)} = (\Lambda_0, MH_{int}\Omega\{\sigma\}). \qquad (2.84)$$

The method under consideration, of cut-off functions in three-dimensional formation, has several advantages as compared with the more consistent four-dimensional formalism, which will be considered in § 3, 4, 5. Thus, e. g., the boundary conditions have a simple physical meaning, the calculations are less complicated, and in several problems the angle variables can be separated. However, along with the above-mentioned advantages, the method entails a number of serious difficulties, connected with the non-renormalizability of terms of the eigenenergy type of the meson and the nucleon, due to the non-covariance of these expressions. In addition, difficulties arise in the Tamm-Dancoff method when terms which describe the polarization of the vacuum are considered. These arise from the fact that the operator of the interaction energy (2.47) contains the product of three field operators, and therefore processes are possible in which three particles are simultaneously created or annihilated. In the actual calculation of the cross section for the π-meson scattering, the eigenenergy terms were altogether omitted /21, 4/ since there are no known approximation methods of calculating these expressions. Although the values of the phase shifts for states with isotopic spin $T = 3/2$, obtained in that calculation, are in a good agreement with experiment, such a simple cutting off of several terms cannot nevertheless be justified from a theoretical point of view. It should also be noted that even if we succeeded in renormalizing the kernel of the integral equation for the scattering, the difficulties due to the eigenenergy of the nucleon would nevertheless arise, if we take into account that the "finite" meson emitted may in reality be again absorbed by the nucleon, if we consider a state with $T = \frac{1}{2}$, $j = \frac{1}{2}$. In solving the integral equation it will then be necessary to integrate over the momentum of this meson.

Additional divergences arise in the method of cut-off equations when we pass to momentum representation /19/, where the infinite expressions cannot be eliminated by renormalization. These infinities arise on account of the fact that in the transition from the differential equations (2.55) to the integral equations, one of the limits of integration remains finite, as distinct from the scattering matrix where the integration is performed between infinite limits. In general, the calculation of the mathematical expectation of the field operators leads to infinite values if we try to determine exactly the time boundaries. For the removal of these new infinities, Stueckellberg has proposed the introduction of a "diffuse" integration limit by replacing discontinuous functions by smooth ones. This method, however, is not a logical outcome of the theory.

The method of cut-off equations has, subsequently, been markedly improved by Dyson. This method, which we are now going to consider, bears the name of the New Tamm-Dancoff method.

§ 3. The generating functional for amplitudes of the new Tamm-Dankov method

As was shown in § 1, the sequence of probability amplitudes (2.32) gives a complete description of the quantum properties of the field. These amplitudes are determined by the operators of non-interacting fields on the one hand, and the vacuum of non-interacting fields* on the other. The concept of mathematical vacuum as a state without particles does not correspond to the real vacuum, since in the latter a continuous creation and annihilation of virtual quanta, which are in a "dynamical equilibrium" with the background, takes place as a result of the interaction of the fields with each other. Dyson has proposed a modification of the theory, in which the state of a system is determined in relation to the real physical vacuum, and not in relation to the mathematical one /22/. The creation and annihilation operators are related, as before to the non-interacting fields. Thus, the vacuum state is a physical one, while the operators remain "mathematical". A consistent generalization of the theory would consist in the introduction of field operators which are subject to non-homogeneous field equations, but in this case it is already impossible to introduce covariantly the creation and annihilation operators by means of which the probability amplitudes ought to be determined. Since the application of the annihilation operator to the physical vacuum state leads to a result different from zero, amplitudes will inevitably arise in the theory, which contain creation operators as well as annihilation operators. These amplitudes are called "the amplitudes" of the new Tamm-Dancoff method and are introduced as follows:

$$a(N, N') = \frac{1}{\sqrt{\Pi(N)\,\Pi(N')}}\,(\Omega_0^{\Phi},\ C(N)\,A(N')\,\Omega) \tag{3.1}$$

In expression (3.1) the following notation has been introduced, following Dyson: the product of all annihilation operators is denoted by the single letter $A(N')$, the product of all creation operators by $C(N)$, where N and N' replace the triplets m, n, l and m', n', l'. Ω_0^{Φ} denotes the physical vacuum state, Ω the vector of the given state, and $\Pi(N) = N!$ In this notation, formulas (2.30) and (2.32) are symbolically written in the form:

$$\psi(N) = \frac{1}{\sqrt{\Pi(N)}}\,(\Lambda_0,\ A(N)\,\Omega),\quad \Lambda(N) = \frac{1}{\sqrt{\Pi(N)}}\,A(N)\,\Lambda_0. \tag{3.2}$$

---

* The mathematical vacuum.

Let us consider a system consisting of one particle and of the physical vacuum. As was mentioned above this system has an undetermined number of particles (an undetermined number of virtual quanta) and therefore it cannot be described by any single probability amplitude; it must be described by the full set of various amplitudes. This circumstance is expressed by the fact that the amplitudes of the new method can be represented as a linear combination of the amplitudes of the old method.

Let us consider the concrete form of the separate Dyson amplitudes (3.1). For that purpose we rewrite the Fourier expansion of the field operators (1.17) and (1.18) in a somewhat different form:

$$\left.\begin{aligned}\psi_\alpha(x) &= \frac{1}{(2\pi)^{3/2}}\sum_u \int d^3k u_\alpha b_{ku} \exp(ik_\mu x_\mu),\\ \overline{\psi}_\alpha(x) &= \frac{1}{(2\pi)^{3/2}}\sum_u \int d^3k \overline{u}_\alpha \overline{b}_{ku} \exp(-ik_\mu x_\mu).\end{aligned}\right\} \quad (3.3)$$

In the new Tamm-Dancoff method, the one-nucleon amplitude is defined by

$$(\Omega_0^\Phi,\ b_{ku}\Omega) = a_1(ku) \quad (3.4)$$

The quantity b_{ku} represent either an annihilation operator or a creation operator, depending to the sign (positive or negative) of the energy to which the spinor index u corresponds; thus, the expression (3.4) for $a_1(ku)$ will correspond either to the one-proton amplitude or to the so-called "minus-one antiproton amplitude". By a "minus particle" we mean a particle which is absent in the state under consideration, but which is present in the physical–vacuum state. A minus particle constitutes a "hole" of its kind in the vacuum of interacting fields. The two-nucleon amplitude will be of the following form

$$(\Omega_0^\Phi,\ Nb_{pu}d_{qv}\Omega), \quad (3.5)$$

where the sign N denotes the normal product of operators. Other amplitudes are formed analogously.

Equation (3.1) for amplitudes can be obtained as follows: We denote by E and E_0 the energy of the given state and of the physical vacuum:

$$(H_0+H_1)\Omega = E\Omega;\quad (H_0+H_1)\Omega_0^\Phi = E_0\Omega_0^\Phi, \quad (3.6)$$

where H_1 is the interaction operator. We have:

$$\begin{aligned}(\Omega_0^\Phi, C(N)A(N')(H_0+H_1)\Omega) &= E(\Omega_0^\Phi, C(N)A(N')\Omega) =\\ &= (\Omega_0^\Phi, (H_0+H_1)C(N)A(N')\Omega) + (\Omega_0^\Phi, [C(N)A(N'), H_0+H_1]\Omega).\end{aligned}$$

The commutator $[C(N)A(N')H_0]$ can be readily computed by means of the relation

$$[c(\mathbf{k}), H_0] = E(k)\,c(\mathbf{k}),$$
$$[c^+(\mathbf{k}), H_0] = -E(k)\,c^+(\mathbf{k}).$$

Denoting $\sum^{N} E(k) = E_N$, $\sum^{N'} E(k) = E_{N'}$, we have

$$(\varepsilon + E_N - E_{N'})\,a(N, N') = \frac{1}{\sqrt{\Pi(N)\,\Pi(N')}}(\Omega_0^{\dagger}, [C(N)\,A(N'), H_1]\,\Omega). \qquad (3.7)$$

The quantity ε, equal to $\varepsilon \equiv E - E_0$, is already a finite quantity, although E and E_0 themselves are infinite. This circumstance constitutes the favorable aspect of the theory. Owing to the fact that on the right-hand side of equation (3.7) we have a commutator, the number of operators on the right-hand side differs from the number of operators on the left-hand side by unity. The amplitude a(N, N') is therefore "linked" with the amplitudes in which the number of particles is greater or less by unity. For these amplitudes, similar equations can be written, which leads to the formation of an infinite system of "linked equations".

There are no vacuum loops in the new method, since the simultaneous creation or annihilation of three particles is impossible in this method /4, 23/. In coordinate space the amplitudes of the new method can be written in the form

$$(\Omega_0^{\dagger}, N\psi(x_1)\ldots\ \psi(x_i), \bar{\psi}(y_1)\ \ldots\ \psi(y_m)\varphi(x_1)\ \ldots\ \varphi(x_n)\,\Omega\{\sigma\}). \qquad (3.8)$$

The state of the system in (3.8) can be given on the hypersurface $\sigma(x)$. The equations for the amplitudes (3.8) can be easily obtained by using the Tomonoga-Schwinger equation for the state vector Ω.

In the new Tamm-Dancoff method the renormalization in the two-nucleon interaction can be performed entirely consistently in the lowest approximation, whereas in the scattering problem difficulties arise, although the expression for the eigenenergy of the nucleon has a covariant form, as distinct from its form in the old method. It should also be noted that difficulties arise in the theory in connection with the appearance of "fictitious poles" of Green's function, which apparently constitutes an overall shortcoming of the quantum theory of fields.

The description of the quantum properties of the field can be obtained in the new Tamm-Dancoff method, just as in the old one, through the introduction of a generating functional for the amplitudes (3.1); it can be written in the form

$$F\{\bar{c}^* c\} = \left(\Omega_0^\Phi, R^+ \exp\left(\int c^*(\mathbf{k})\, c^+(\mathbf{k})\, d^3k\right) R \exp\left(\int \bar{c}(\mathbf{k})\, c(\mathbf{k})\, d^3k \quad \Omega\right)\right) \tag{3.9}$$

where R is defined by (2.40).

Let us establish the connection between the generating functionals of the new (3.9) and of the old (2.41) method. For the sake of brevity we consider only a Bose field. Expression (3.9) can be represented in the form

$$F\{\bar{c}^*\bar{c}\} = \sum_{n=0}^{\infty} \left(\Omega_0^\Phi, \exp\left(\int c^*(\mathbf{k})\, c^+(\mathbf{k})\, d^3k\right) \Lambda_n\right) \times \\ \times \left(\Lambda_n, \exp\left(\int \bar{c}(\mathbf{k})\, c(\mathbf{k})\, d^3k\right) \Omega\right) \tag{3.10}$$

where Λ_n, the quantities defined by (2.20), constitute a complete orthonormal system of functions. The second co-factor of (3.10) can be written as follows:

$$\left(\Lambda_n, \exp\left(\int c(\mathbf{k})\, c(\mathbf{k})\, d^3k\right) \Omega\right) = \frac{\delta^n}{\delta \bar{c}^n(\mathbf{k})} \frac{1}{\sqrt{n!}} \times \\ \times \left(\Lambda_0, \exp\left(\int \bar{c}(\mathbf{k})\, c(\mathbf{k})\, d^3k\right) \Omega\right) \tag{3.11}$$

From (3.11) follows

$$F = \left(\Omega_0^\Phi, \exp\left(\int (\bar{c}^*(\mathbf{k}) + \delta/\delta\,\bar{c}(\mathbf{k}))\, c^+(\mathbf{k})\, d^3k\right) \Lambda_0\right) \times \\ \times \left(\Lambda_0, \exp\left(\int c(\mathbf{k})\, c(k)\, d^3k\right) \Omega\right) = \\ = \left(\Omega_0^\Phi, \exp\left(\int \left(c^*(\mathbf{k}) + \delta/\delta c(\mathbf{k})\right) c^+(\mathbf{k})\, d^3k\right) \Lambda_0\right) \tilde{\Omega}(\bar{c}) \tag{3.12}$$

Expression (3.12) establishes in fact the connection between the above-mentioned functionals /12/. Here $\tilde{\Omega}(\bar{c})$ is given by formula (2.41).

II. The Generating Functionals for Relativistic Functions and Functional Integration

The method of Fock functionals constitutes one of the possible rigorous functional formulations of the quantum theory of fields. In this method, Fock developed the basic idea of the generating functional by choosing as the fundamental functions which describe the field, an infinite sequence of probability amplitudes in configuration space. It is clear that we obtain another rigorous functional formulation of the quantum theory of fields if we develop this same idea of generating functional on the basis of a complete sequence of some other functions which depend on variables of a definite number of particles. This could be seen already when considering, in § 3, the generating functional for Dyson amplitudes (the amplitudes of the new Tamm-Dancoff method). It is characteristic of the Fock functional as well as of the functional of the new Tamm-Dancoff method, that they depend on functions of a vector argument or, in general, on functions of a space-time point, lying on a space-like surface.

We shall consider in § 4 the development of the idea of the generating functional in the case where the field is described by a set of relativistic functions, either T-functions, or Feynman-amplitudes, or ρ-functions.

In relativistic functions, the space—time variables of the particles can be arbitrary and therefore the generating functionals will, in that case, be functionals of functions of a space-time point, which have the meaning of external field-sources. The functionals of external sources were introduced by Schwinger /27/. The equations for the Schwinger generating-functional and for the above-mentioned functions are simply deduced from the equations for the field operators in the Heisenberg representation. The equations for the Feynman amplitudes may be considered /28/ as a four-dimensional generalization of the Fock equations for probability amplitudes.

The Feynman amplitudes and the T-functions are directly related to the amplitude of the transition between states for $t = \pm\infty$. The description of the field by means of these functions signifies a space-time approach to field theory. The space-time treatment of the field theory (initially proposed by Feynman /29/), will, apparently, also be preserved in the future theory, which will be free of the contradictions of the present-day theory. Indeed, so far as the corpuscular aspect will be preserved in the future theory, the fundamental quantities which characterize the field state will be as before, relativistic functions of the Feynman-amplitude type /31/ or generating functionals for these. Besides, a number of considerations favor the fact that, in the space-time treatment where development in time is not considered and the canonical formalism is lacking, it will be easier to formulate a theory which does not use the concept of "bare" particles. On the grounds a consistent attempt to construct, from the very foundations, a "four-dimensional" formalism which is to replace the "three-dimensional" treatment, has been undertaken by several authors /34, 35/. § 5 of the present survey is devoted to the space-time treatment. In the same paragraph, Fourier functional transformations are considered and by their means the general solution of the problem of interacting fields is found in the form of a continuous integral over the Bose and Fermi fields.

Integration over a Fermi field, i.e., over anticommuting functions, requires great care. The method of integration over a Fermi field was indicated by Matthews and Salam /36/. In § 6 of our survey we shall, following these authors, express the variation of the operator of the Fermi field in terms of the variation of ordinary numbers. Owing to this, it is possible to put the equation for the four-state vector into a form which has no anticommuting functional derivatives.

Approximation methods for functional integration and renormalization problems are not considered.

§ 4. The generating functionals for relativistic functions

1. The T-function* and the generating functional. The probability amplitudes Ψ_{nml} in coordinate space

$$\Psi_{nml}(x\ldots|y\ldots|z\ldots)=(\Lambda_0,\psi_0^{(+)}x\ldots\overline{\psi}_0^{(+)}(y)\ldots\varphi_0^{(+)}(z)\ldots\Omega(\sigma)) \tag{4.1}$$

are not "four-dimensional" functions, since the coordinates x... y... z... in (4.1) ought to lie on some space-like surface σ. In addition, as has been explained in § 2, 4, calculations with probability amplitudes Ψ_{nml} present many difficulties which are ultimately due to the fact that the amplitudes Ψ_{nml} represent the expansion coefficients of the state vector $\Omega(\sigma)$ in terms of the eigenfunctionals of the operators of the number of particles for non-interacting fields. Calculations with amplitudes of the Dyson type (§ 3) have also their peculiar difficulties, arising from their inconsistent definition in which the operators of noninteracting fields operate on the vacuum of interacting fields.

It is convenient to use the Heisenberg representation for the definition of the relativistic wave functions. In the Heisenberg representation, the field operators $\psi(x)$, $\overline{\psi}(x)$ and $\varphi(x)$ satsify the equations:

$$D(x)\psi(x)-g\gamma_5\varphi(x)\psi(x)=0, \tag{4.2}$$

$$D(-\overline{x})\psi(x)-g\gamma_5(x)\varphi(\overline{x})\psi(x)=0, \tag{4.3}$$

$$K(x)\varphi(x)-g\frac{1}{2}[\overline{\psi}(x),\gamma_5\psi(x)]=0, \tag{4.4}$$

where $D(x)=i\left(\gamma_\lambda(x)\frac{\partial}{\partial x_\lambda}+m\right)$, $K(x)=i(\Box_x-\mu^2)$, and where $\gamma_\lambda(x)$ means that the matrix γ_λ is multiplied by a spinor which depends on x: $\gamma_\lambda(x)\overline{\psi}(x)=\overline{\psi}(x)\gamma_5$, $\gamma_\lambda(x)\overline{\psi}(y)\ldots\psi(x)=\psi(y)\ldots\gamma_\lambda\overline{\psi}(x)$.

In the Heisenberg representation, the amplitudes cannot be introduced by formulas of type (4.1), but by means of Heisenberg operators and state vectors, since the Heisenberg operators ψ, $\overline{\psi}$ and φ cannot be

---

* [Translator's Note: on R. page 53, in the Contents, appears "T-functions". The ingular is used here in the generic sense, as descriptive of the class of functions

subdivided, in a relativistic-invariant manner, into positive- and negative-frequency parts. The relativistic functions of simplest form are the matrix elements of chronological T-products of the field operators themselves*:

$$T_{nml}(x_1 \ldots x_n \mid y_1 \ldots y_m \mid z_1 \ldots z_l) = \\ = (\Psi_0, T[\psi(x_1) \ldots \psi(x_n)\overline{\psi}(y_1) \ldots \overline{\psi}(y_m)\varphi(z_1) \ldots \varphi(z_l)]\,\Psi), \qquad (4.5)$$

where Ψ is the state vector in the Heisenberg representation, and Ψ_0 refers to the physical vacuum; the functions $T_{nml}(x \ldots | y \ldots | z \ldots)$ are symmetric in the variables z and antisymmetric in the variables x...y...

In the functions $T_{nml}(x \ldots | y \ldots | z \ldots)$, the coordinates and the time of the points x... y... z... are not restricted by the requirement that they should lie on some space-like surface.

By means of the functions T_{nml} it is possible to find the energy of stationary states and the matrix elements of the transitions. Let Ψ_a describe the state which belongs to the eigenvalue $p_\mu^{(a)}$ of the energy-momentum operator P_μ of interacting fields:

$$P_\mu \Psi_a = p_\mu^{(a)} \Psi_a. \qquad (4.6)$$

It is then possible to determine the eigenvalues $p_\mu^{(a)}$ from the equation

$$\begin{aligned} p_\mu^{(a)} T_{nml}(x \ldots | y \ldots | z \ldots) &= (\Psi_0, T[\psi(x) \ldots \overline{\psi}(y) \ldots \varphi(z) \ldots] P_\mu \Psi_a) = \\ &= \left(\Psi_0, T\left[-i\frac{\partial \psi(x)}{\partial x_\mu} \ldots \overline{\psi}(y) \ldots \varphi(z)\right]\Psi_a\right) + \ldots \\ &\ldots + \left(\Psi_0, T\left[\psi(x) \ldots -i\frac{\partial \overline{\psi}(y)}{\partial y_\mu} \ldots \varphi(z)\right]\Psi_a\right) + \ldots \\ &\ldots + \left(\Psi_0, T\left[\psi(x) \ldots \overline{\psi}(y) \ldots -i\frac{\partial \varphi(z)}{\partial z_\mu}\right]\Psi_a\right) = \\ &= -i\sum_{nml}\left(\frac{\partial}{\partial x_\mu} + \ldots + \frac{\partial}{\partial y_\mu} + \ldots + \frac{\partial}{\partial z_\mu}\right) T_{nml}(x \ldots | y \ldots | z \ldots). \end{aligned} \qquad (4.7)$$

Let us consider the relation between the functions $T_{nml}(x \ldots | y \ldots | z \ldots)$ and the transition amplitudes in the case of unrelated states and when,

---

* In a chronological product the order of the operators is such that the time at which the operators are taken increase from right to left. A T-product differs from a chronological one by the factor $(-1)^n$, where n is the number of permutations of the operators of a Fermi field, which are required for the chronological ordering of the product /41/.

for $t=\pm\infty$ only free particles are present. Then, for $t\to-\infty$ the field operators $\psi, \bar{\psi}$ asymptotically change into the operators $\psi_{in}, \bar{\psi}_{in}$, and for $t\to+\infty$ into the operators $\psi_{out}, \bar{\psi}_{out}$, which describe the free particles. If Ψ^{in}_{α} and $\Psi^{out}_{\alpha'}$ are the state vectors in the Heisenberg representation, defined by means of the "in" and "out" field-operators, then the transition amplitude $U_{\alpha\alpha'}$ is connected with the matrix element of the S - matrix by the relation

$$U_{\alpha\alpha'}=(\Psi^{out}_{\alpha'}, \Psi^{in}_{\alpha})=(\Psi^{in}_{\alpha'}, S\Psi^{in}_{\alpha}). \tag{4.8}$$

Since $\Psi^{out}_{\alpha'}$ and Ψ^{in}_{α} describe a field state with free particles, we can use formula (2.30) for the representation of Ψ^{in}_{α} writing it in the form

$$\Psi^{in}_{nml}=(n!\,m!\,l!)^{-1/2}\,a^{+}_{in}(\mathbf{p}_1)\ldots a^{+}_{in}(\mathbf{p}_n)\,b^{+}_{in}(\mathbf{q}_1)\ldots b^{+}_{in}(\mathbf{q}_n)\,c^{+}_{in}(\mathbf{k}_1)\ldots c^{+}_{in}(\mathbf{k}_b)\,\Psi_0 \tag{4.9}$$

We proceed analogously for Ψ^{out}_{nml}. Let us discuss this on the example of meson scattering on nucleons. Then

$$\Psi^{in}_{\alpha}=\Psi^{in}_{101}=a^{+}_{in}(\mathbf{p})\,c^{+}_{in}(\mathbf{k})\Psi_0,\ \Psi^{out}_{\alpha'}=\Psi^{out}_{101}=a^{+}_{ont}(\mathbf{p}')\,c^{+}_{out}(\mathbf{k}')\,\Psi_0.$$

Substituting in (4.8) the expressions for the creation operators a^+ and c^+ obtained from (1.13) and (1.14), we find:

$$\begin{aligned}U(\mathbf{pk};\,\mathbf{p}'\mathbf{k}')=\int f_{p'k'}(xz')\,i\,\frac{\partial}{\partial z'_0}\,d^3z'\gamma_4(x)\,d^3x\,(\Psi_0,\,\varphi(z')\,\psi(x)\,\bar{\psi}(y)\,\varphi(z)\,\Psi_0)\times\\ \times i\,\frac{\partial}{\partial z_0}\,d^3z\gamma_4(y)\,f_{pk}(yz),\end{aligned} \tag{4.10}$$

$$x_0,\,z'_0\to+\infty;\quad y_0,\,z_0\to-\infty,$$

where $f_{pk}(yz)$ and $f_{p'k'}(xz')$ are the wave functions of the incident and scattered particles, which we assume to be orthogonal. The mean vacuum value of the field operators in (4.10) can be expressed by the vacuum function $T^{\circ}_{112}\equiv\tau^{\circ}_{112}$:

$$(\Psi_0,\,\varphi(z')\,\psi(x)\,\bar{\psi}(y)\,\varphi(z)\,\Psi_0)=T^{\circ}_{112}(x\,|\,y\,|\,z'z)\equiv\tau^{\circ}_{112}(x\,|\,y\,|\,zz'), \tag{4.11}$$

since in (4.10) x_0, $z'_0\to+\infty$; y_0, $z_0\to-\infty$. The three-dimensional integrals in (4.10) can be transformed into four-dimensional ones, if we introduce into (4.10) the function τ_{112} from (4.11), because in that case, for x_0, $z'_0\to-\infty$ or y_0 , $z_0\to+\infty$, the quantity $U(\mathbf{pk};\,\mathbf{p}'\mathbf{k}')$ vanishes by definition of the vacuum. After the transformation, expression (4.10) becomes

$$U(p,k;\, p'k') = \int \bar{f}_{p'k'}(xz')\, \{K(z')\,K(z)\,D(-y)\,D(x)\,\tau_{112}(x\,|\,y\,|\,zz')\} \times \\ \times f_{pk}(yz)\, d^4x\,d^4y\,d^4z\,d^4z'.$$

In the general case we obtain, for different states $\Psi^{out}_{\alpha'}$ and Ψ^{in}_{α}:

$$\begin{aligned} U_{\alpha\alpha'} = (-1)^{\alpha+\alpha'} \int \bar{f}_{\alpha'}(y'\ldots|x'\ldots|z'\ldots)\, \{D(-y')\ldots D(+x)\ldots K(z')\ldots \\ \ldots D(x)\ldots D(-y)\ldots K(z)\ldots \times \tau_{\alpha+\alpha'}(x'\ldots x\ldots|y'\ldots y\ldots|z'\ldots z\ldots)\} \times \\ \times f_{\alpha}(x\ldots|y\ldots|z\ldots)\, d^4x\ldots d^4x'\ldots d^4y\ldots d^4y'\ldots d^4z\ldots d^4z'\ldots, \end{aligned} \tag{4.12}$$

where f_α and $f_{\alpha'}$ are the wave functions of the incident and scattered particles. Thus, for the calculation of the transition amplitudes, it is sufficient to know the vacuum functions $\tau(x\ldots|y\ldots|z\ldots)$ only /32, 39, 40/.

For the description of the field it is necessary to know, in the general case, the entire set of functions $T_{nml}(x\ldots|y\ldots|z\ldots)$; $n, m, l = 0, 1, 2, \ldots \infty$. In the same manner as we may consider, instead of the set of probability amplitudes $\Psi_{nml}(p\ldots|q\ldots|k\ldots)$, the generating functional $\Omega\{a, b, c\}$ (§ 2, 3), we may likewise consider, instead of the entire set of functions T_{nml}, the corresponding generating functional /32, 47, 50/. Since the functions $T_{nml}(x\ldots\, y\ldots\, z\ldots)$ depend on space-time coordinates, it is necessary to introduce, for the construction of the generating functional, the auxiliary quantities $\eta(x), \bar{\eta}(y)$ and the function $I(z)$, which also depend on space-time coordinates and which are proportional to the unitary matrix in the occupation-number space for the nucleon and meson fields. It is convenient to write this functional $Z\{\eta, \bar{\eta}, I\}$ in the form

$$\left.\begin{aligned} Z\{\eta, \bar{\eta}, I\} = \sum_{nml}^{\infty} \frac{i^{n+m+l}}{n!\,m!\,l!} Z_{nml}\{\eta, \bar{\eta}, I\}, \\ Z_{nml}\{\eta, \bar{\eta}, I\} = \\ = \int \bar{\eta}(x_n)\ldots\bar{\eta}(x_1)\, T_{nml}(x_1\ldots x_n\,|\,y_1\ldots y_m\,|\,z_1\ldots z_l) \times \\ \times \eta(y_m)\ldots\eta(y_1)\, I(z_1)\ldots I(z_l)\, d^4x_1\ldots d^4x_n \times \\ \times d^4y_1\ldots d^4y_m d^4z_1\ldots d^4z_l. \end{aligned}\right\} \tag{4.13}$$

From the antisymmetry of $T_{nml}(x\ldots|y\ldots|z\ldots)$ with respect to the variables $x\ldots, y\ldots,$ it follows, moreover, that the quantities $\eta(y), \bar{\eta}(x)$, commute with each other:

$$\left.\begin{aligned} \{\eta(x), \eta(x')\} = 0, \quad \{\eta(x), \bar{\eta}(y)\} = 0, \\ \{\bar{\eta}(y), \bar{\eta}(y')\} = 0, \quad [\eta(x), I(x')] = 0, \end{aligned}\right\} \tag{4.14}$$

hence it follows that η and $\bar{\eta}$ are proportional to the external sources of the nucleon field (see § 2, 3). The factors in expansion (4.13) have been chosen in such a way, that η, $\bar{\eta}$ and I have the meaning of external sources. Indeed, from definition (4.5) of the functions T_{nml} follows that (4.13) can be symbolically represented as a series expansion of the quantity

$$Z\{\eta, \bar{\eta}, I\} = (\Psi_0, \tau\{\eta, \bar{\eta}, I\}\Psi), \tag{4.15}$$

where the operator $\tau\{\eta, \bar{\eta}, I\}$ is given by the expression

$$\tau\{\eta, \bar{\eta}, I\} = T\exp\left\{i\int[\bar{\eta}(x)\psi(x) + \bar{\psi}(x)\eta(x) + I(x)\varphi(x)]\,d^4x\right\} =$$
$$= T\exp\left[-i\int_{-\infty}^{\infty} H'(t)\,dt\right]$$

and where H'(t) represents that part of the interaction energy operator which depends on the external sources. If we consider $\tau\{\eta, \bar{\eta}, I\}$ as a limit, for $t \to +\infty$, of the operator

$$\tau\{\eta, \bar{\eta}, I; t\} = T\exp\left[-i\int_{-\infty}^{t} H'(t')\,dt'\right], \tag{4.16}$$

then the state vector $\Psi' = \tau\{\eta, \bar{\eta}, I; t\}\Psi$ will change in time, by virtue of its interaction with the external sources

$$i\frac{\partial\Psi'}{\partial t} = H'\Psi'. \tag{4.17}$$

In all these formulas we assume that the operators ψ, $\bar{\psi}$ and φ do not depend on the external sources (see equations (4.2)-(4.4)). If the functional $Z\{\eta, \bar{\eta}, I\}$ is known, then the function

$$T_{nml}(x\ldots|y\ldots|z\ldots)$$

is expressed in terms of the (n + m + l)-th functional derivative of $Z\{\eta, \bar{\eta}, I\}$:

$$T_{nml}(x_1\ldots x_n|y_1\ldots y_m|z_1\ldots z_l) =$$
$$= (-i)^{n+m+l}\left.\frac{\delta^{n+m+l}Z\{\eta, \bar{\eta}, I\}}{\delta\bar{\eta}(x_1)\ldots\delta\bar{\eta}(x_n)\,\delta\eta(y_1)\ldots\delta\eta(y_m)\,\delta I(z_1)\ldots\delta I(z_l)}\right|_{\eta=\bar{\eta}=I=0}, \tag{4.18}$$

where we have to put, in the result, $\eta = 0$, $\bar{\eta} = 0$ and $I = 0$. By means of (4.18) it is possible to obtain the following useful relations: If $Z'\{\eta, \bar{\eta}, I\} = \eta(\xi)Z\{\eta, \bar{\eta}, I\}$, then for the corresponding functions T'_{nml} and T_{nml} the following equation holds

$$T'_{nml}(x_1..|y_1...y_m|z...) = -i\sum^m \delta^4(\xi - y_1)T_{nm-1l}(x...|y_2...y_m|z...), \qquad (4.19)$$

whose right-hand side is antisymmetric in the variables y. If $Z''\{\eta, \bar{\eta}, I\} = \frac{\delta}{\delta\bar{\eta}(\xi)} Z\{\eta, \bar{\eta}, I\}$, then the functions T''_{nml} and T_{nml} are connected by relation

$$T''_{nml}(x_1...x_n|y...|z...) = iT_{n+1ml}(\xi x_1...x_n|y...|z...)(-1)^{n+m}. \qquad (4.20)$$

If $Z'''\{\eta, \bar{\eta}, I\} = \frac{\delta}{\delta J(z)} Z\{\eta, \bar{\eta}, I\}$, then we obtain for the functions T'''_{nml} and T_{nml} the equation

$$T'''_{nml}(x...|y...|z_1...z_l) = iT_{nml+1}(x...|y...|zz_1...z_l) \text{ etc.} \qquad (4.21)$$

The equations for the generating functional $Z\{\eta, \bar{\eta}, I\}$ can be obtained from the equations for the functions $T_{nml}(x...|y...|z...)$. In the simplest case, we obtain the functions $T_{110}(x|y|-) = (\Psi_0, T[\psi(x)\bar{\psi}(y)]\Psi)$ from equation (4.2) for the operator $\Psi(x)$ and the definition of the T-products:

$$\begin{aligned} D(x)T_{110}(x|y|-) &= D(x)(\Psi_0, \tfrac{1}{2}[\psi(x), \bar{\psi}(y)]\Psi) + \\ &+ D(x)\varepsilon(x, y)\left(\Psi_0, \tfrac{1}{2}\{\psi(x), \bar{\psi}(y)\}\Psi\right) = \\ &= g\gamma_5(x)(\Psi_0, T[\psi(x)\bar{\psi}(y)\varphi(x)]\Psi) + \delta^4(x-y), \end{aligned}$$

since $\partial\varepsilon(x)/\partial x_0 = 2\delta(x_0)$ and since, for equal times $x_0 = y_0$, the commutator $\{\psi(x), \bar{\psi}(y)\} = \gamma_4\delta^3(x-y)$. Thus $T_{110}(x|y|-)$ satisfies equation

$$D(x)T_{110}(x|y|-) = g\gamma_5(x)T(x|y|x) + \delta^4(x-y).$$

In the general case, the chronological product of the operator $\psi(x)$ and of n other operators $A(x_1)... A(x_n)$ (A may be equal to ψ, $\bar{\psi}$ or φ) can be represented in the form

$$\begin{aligned} T[\psi(x), A(x_1), \ldots, A(x_n)] = \Sigma(-1)^P\psi(x)T[A(x_1), \ldots, A(x_n)] \times \\ \times\theta(x-x_1), \ldots, \theta(x-x_n), \end{aligned} \qquad (4.22)$$

where the sum is taken over all the permutations of x and x_k; P is equal to the number of permutations of $\psi(x)$ with the operators $\psi(x_k)$ and $\bar{\psi}(x_l)$; $\theta(x) = \frac{1}{2}[1 + \varepsilon(x)]$. Since for equal times $x_0 = y_0$, we shall have $\{\psi(x), \bar{\psi}(y)\} = [\psi(x), \varphi(y)] = 0$, then, by (4.2) and (4.22), the function

$T_{nml}(x\ldots|y\ldots|z\ldots)$ satisfies the following equation with respect to the coordinate x_1:

$$D(x_1)T_{nml}(x_1\ldots|y_1\ldots|z_1\ldots)=g\gamma_5(x_1)T_{nml+1}(x_1\ldots|y_1\ldots|x_1z_1\ldots)+$$
$$+\sum_i^m \delta^4(x_1-y_i)T_{n-1m-1l}(x_2\ldots|y_2\ldots|z_1\ldots). \tag{4.23}$$

From expression (4.18) for the functions T_{nml} and from formulas (4.19)-(4.21), it follows that equation (4.23) can also be written in the form

$$\left\{D(x)\frac{\delta}{\delta\bar{\eta}(x_1)}-ig\gamma_5\frac{\delta}{\delta\bar{\eta}(x_1)}\frac{\delta}{\delta I(x_1)}+\eta(x_1)\right\}\times$$
$$\times\frac{\delta^{n+m+l-1}Z\{\eta,\bar{\eta},I\}}{\delta\bar{\eta}(x_1)\ldots\delta\eta(y_1)\ldots\delta I(z_1)\ldots}=0 \tag{4.24}$$

for $\eta=\bar{\eta}=I=0$.

In view of the arbitrariness of the numbers n, m, l, this equation will hold if $Z\{\eta,\bar{\eta},I\}$ satisfies the equation

$$\left\{D(x)\frac{\delta}{\delta\bar{\eta}(x)}-ig\gamma_5\frac{\delta}{\delta\bar{\eta}(x)}\frac{\delta}{\delta I(x)}\right\}Z\{\eta,\bar{\eta},I\}=-\eta(x)Z\{\eta,\bar{\eta},I\}. \tag{4.25}$$

This is Schwinger's first-variation equation for the generating functional $Z\{\eta,\bar{\eta},I\}$. Two other variation equations can be obtained in the same way from the equations for $T_{nml}(x\ldots|y\ldots|z\ldots)$ with respect to the variables y... and z..., which are, in turn, equivalent to the operator equations (4.3) and (4.4). We have:

$$\left\{D(-y)-ig\gamma_5(y)\frac{\delta}{\delta I(y)}\right\}\frac{\delta}{\delta\eta(y)}Z\{\eta,\bar{\eta},I\}=\bar{\eta}(y)Z\{\eta,\bar{\eta},I\} \tag{4.26}$$

and

$$K(z)\frac{\delta Z\{\eta,\bar{\eta},I\}}{\delta I(z)}=\left\{I(z)+ig\frac{\delta}{\delta\eta(z)}\gamma_5\frac{\delta}{\delta\bar{\eta}(z)}\right\}Z\{\eta,\bar{\eta},I\}. \tag{4.27}$$

It is necessary to impose boundary conditions on the system of equations (4.25), (4.26), (4.27). If, e.g., $\Psi=\Psi_a$ refers to a state with energy momentum $p_\mu^{(a)}$, $a\neq 0$, then, for $\eta=\bar{\eta}=I=0$ one of the boundary conditions is of the form

$$Z\{0,0,0\}=0; \tag{4.28a}$$

whereas if $\Psi=\Psi_0$ then, by (4.15) we should have

$$Z\{0,0,0\}=1. \tag{4.28b}$$

The other boundary conditions are also determined by the meaning of the state Ψ; if, e. g., $\Psi = \Psi_0$ is the physical vacuum, we should have, for $\eta = \bar{\eta} = I = 0$

$$\frac{\delta Z}{\delta \eta} = \frac{\delta Z}{\delta \bar{\eta}} = \frac{\delta Z}{\delta I} = 0 \tag{4.29}$$

The equations with functional derivatives (4.26), (4.27) and (4.28) for $Z\{\eta, \bar{\eta}, I\}$ can be solved in the general form /42-46/. This constitutes the advantage of the method of functionals, for the consistent application of other methods leads, in one way or another, to the use of the perturbation theory.

The solution of the equations for $Z\{\eta, \bar{\eta}, I\}$ and related problems will be considered in § 5 and 6.

2. Feynman amplitudes and the generating functional. Let us consider the problem of constructing relativistic wave functions, which are the four-dimensional analogues of the ordinary wave functions of a system of particles (probability amplitudes) in configuration space $\psi_{nml}(x\ldots|y\ldots\ldots|z\ldots)$ (formula (4.1). The direct generalization of definition (4.1) to the case of Heisenberg operators is impossible since, as we already noted in § 4, 1, the Heisenberg operators cannot be subdivided, in an invariant way, into positive- and negative frequency parts. In any case, the four-dimensional wave functions $f_{nml}(x\ldots|y\ldots|z\ldots)$ ought to possess the following properties:

a) The functions $f(x\ldots|y\ldots|z\ldots)$ should be antisymmetric in the nucleon and anti-nucleon variables $x\ldots$, $y\ldots$ and symmetric in the meson variables $z\ldots$

b) The equations for the functions $f_{nml}(x\ldots|y\ldots|z\ldots)$ should not have singularities for the case of coinciding variables (or at least for coinciding times) of any two particles; the interval between two points x and y can be arbitrary.

c) In the interaction, the function $f_{nml}(x\ldots|y\ldots|z\ldots)$ should be, to a constant factor, equal to the probability amplitude $\psi_{nml}(x\ldots|y\ldots|z\ldots)$

d) For stationary states the following relation should hold:

$$-i\sum^{n,m,l}\left\{\frac{\partial}{\partial x_\mu}+\ldots+\frac{\partial}{\partial y_\mu}+\ldots+\frac{\partial}{\partial z_\mu}+\ldots\right\}f_{nml}(x\ldots|y\ldots|z\ldots)= \\ = p_\mu^{(a)}f_{nml}(x\ldots|y\ldots|z\ldots), \tag{4.30}$$

where $p_\mu^{(a)}$ is the eigenvalue of the energy-momentum vector in the stationary state.

It can be seen from definition (4.5) and equation (4.23), the functions $T_{nml}(x\ldots|y\ldots|z\ldots)$ do not possess the properties b) and c) and therefore cannot serve as four-dimensional wave functions.

If there is no interaction, the relation between the amplitudes Ψ_{nml} and the functions T_{nml} can readily be found by means of Wick's formula for T-products and N products (see, e.g., /41/), since the probability amplitude Ψ_{nml} is nothing else but the matrix element of the N-products:

$$\Psi_{nml}(x\ldots|y\ldots|z\ldots)=(\Lambda_0, N[\psi(x)\ldots\overline{\psi}(y)\ldots\varphi(z)\ldots]\,\Phi), \qquad (4.1')$$

and in the absence of interaction, the intervals between the points x..., ..., y..., z..., can be arbitrary. In view of definitions (4.1') and (4.5) of the functions Ψ_{nml} and T_{nml}, and of Wicks formula, we obtain, that in the absence of interaction

$$\begin{aligned}T^0_{nml}(x\ldots|y\ldots|z\ldots)&=\Psi_{nml}(x\ldots|y\ldots|z\ldots)+\\&+\Sigma\Delta_F(z_l-z_n)\,\Psi_{nml-2}(x\ldots|y\ldots|zz_l^{-1}z_k^{-1})-\\&-\Sigma S_F(x_i-y_k)\,\Psi_{n-1m-1l}(x\ldots x_i^{-1}|y\ldots y_k^{-1}\ldots|z\ldots)+\ldots,\end{aligned} \qquad (4.31)$$

where x_i^{-1} denotes that $\Psi_{n-1ml}(x\ldots x_i^{-1}|y\ldots|z\ldots)$ does not depend on the variable x_i; Δ_F and S_F are the Feynman-Green's functions for free fields* and T^0_{nml} is the T_{nml} function in the absence of interaction.

From (4.31) it is evident that the singularities of the functions $T^0_{nml}(x\ldots|y\ldots|z\ldots)$ have the same character as the singularities of the Green's functions Δ_F and S_F.

If T_{nml} is defined by means of the operators of interacting fields, we may use formula (4.31) as an indication of the character of the relation between T_{nml} and f_{nml}. Conditions a), b), c) and d) do not determine this relation uniquely, since the subtraction of the singularities can be performed by different methods.

If we assume that in the presence of interaction, the singularities

---

* Green's functions $S_F(x-y)=(\Lambda_0, T[\psi(x)\overline{\psi}(y)]\,\Lambda_0)$ and $\Delta_F(x-y)=(\Lambda_0, T[\varphi(x)\varphi(y)]\,\Lambda_0)$ satisfy the equations $D(x)\,S_F(x-y)=\delta^4(x-y)$, $K(x)\,\Delta_F(x-y)=-\delta^4(x-y)$.

of the functions $T_{nml}(x\ldots|y\ldots|z\ldots)$ have the same character as in its absence, then it is possible to define, for interacting fields, the functions $f(x\ldots\, y\ldots\, z\ldots)$ either by replacing in formula (4.31) the functions Ψ_{nml} and T°_{nml} for the free fields, by the functions f_{nml} and T_{nml}:

$$\begin{aligned}T_{nml}(x\ldots|y\ldots|z\ldots) &= f_{nml}(x\ldots|y\ldots|z\ldots)+\\&+\Sigma\Delta_F(z_i-z_k)\,f_{nml-2}(x\ldots|y\ldots|z\ldots z_i^{-1}z_k^{-1}\ldots)-\\&-\Sigma S_F(x_i-y_k)\,f_{n-1m-1l}(x\ldots x_i^{-1}\ldots|y\ldots y_k^{-1}\ldots|z\ldots)+\ldots,\end{aligned} \tag{4.32}$$

or by solving (4.32) with respect to f_{nml}:

$$\begin{aligned}f_{nml}(x\ldots|y\ldots|z\ldots) &= T_{nml}(x\ldots|y\ldots|z\ldots)-\\&-\Sigma\Delta_F(z_i-z_k)\,T_{nml-2}(x\ldots|y\ldots|z\ldots z_i^{-1}\ldots z_k^{-1})+\\&+\Sigma S_F(x_i-y_k)\,T_{n-1m-1l}(x\ldots x_i^{-1}\ldots|y\ldots y_k^{-1}|z\ldots)+\ldots\end{aligned} \tag{4.33}$$

Definition (4.33) satisfies conditions a), b), c), d). The functions $f(x\ldots|y\ldots|z\ldots)$ are defined in such a way in references /28, 40, 49/. However, Lehmann has shown /23/, that the singularities of the Green's functions Δ'_F and S'_F for interacting fields cannot be less than the singularities of Δ_F and S_F. The same conclusion can also be drawn with regard to the character of the singularities of the functions T_{nml}, as compared with the singularities of the functions T^{o}_{nml}. Therefore, generally speaking, definition (4.33) may not satisfy condition b). From this point of view, it is expedient to define the functions $f'(x\ldots|y\ldots|z\ldots)$ by means of the functions Δ'_F and S'_F instead of Δ_F and S_F:

$$\begin{aligned}f'_{nml}(x\ldots|y\ldots|z\ldots) &= T_{nml}(x\ldots|y\ldots|z\ldots)-\\&-\Sigma\Delta'_F(z_i-z_k)\,T_{nml-2}(x\ldots|y\ldots|z\ldots z_i^{-1}\ldots z_k^{-1})+\\&+\Sigma S_F(x_i-y_k)\,T_{n-1m-1l}(x\ldots x_i^{-1}\ldots|y\ldots y_k^{-1}\ldots|z\ldots)-\ldots,\end{aligned} \tag{4.34}$$

which, however, leads to more cumbersome equations for f'_{nml}. The development in time of the functions T_{nml}, Δ_F, S_F, Δ'_F, S'_F takes place in the Feynman form: the waves of positive energy propagate themselves forward in time and the waves of negative energy backwards in time. From definitions (4.32)—(4.34) of the functions $f_{nml}(x\ldots|y\ldots|z\ldots)$ it follows that these functions develop similarly in time. The functions $f_{nml}(x\ldots|y\ldots|z\ldots)$ are sometimes called Feynman amplitudes.

The generating functional $S\{\eta, \bar{\eta}, I\}$ for the four-dimensional wave functions f_{nml}:

$$\left.\begin{aligned} S\{\eta, \bar{\eta}, I\} &= \sum_{nml}^{\infty} \frac{i^{n+m+l}}{n!\, m!\, l!} S_{nml}\{\eta, \bar{\eta}, I\}, \\ S_{nml}\{\eta, \bar{\eta}, I\} &= \int \bar{\eta}(x_n)\ldots\bar{\eta}(x_1) f_{nml}(x_1 \ldots x_n \mid y_1 \ldots y_m \times \\ &\times \mid z_1, \ldots z_l)\, \eta(x)\ldots\eta(x)\, I(z_1)\ldots I(z_l)\, d^4x_1\ldots d^4y_1\ldots d^4z, \end{aligned}\right\} \quad (4.35)$$

is related to the functional of external sources $Z\{\eta, \bar{\eta}, I\}$ by the transformation

$$S\{\eta, \bar{\eta}, I\} = e^{\frac{1}{2} I\Delta_F I + \bar{\eta} S_F \eta} Z\{\eta, \bar{\eta}, I\}, \quad (4.36)$$

where we introduced the notation

$$\left.\begin{aligned} I\Delta_F I &= \int d^4x d^4y I(x)\, \Delta_f(x-y)\, I(y), \\ \bar{\eta} S_F \eta &= \int d^4x d^4y \bar{\eta}(x)\, S_F(x-y)\, \eta(y). \end{aligned}\right\} \quad (4.37)$$

In order to convince ourselves of the validity of formula (4.35), we have to expand both sides of formula (4.36) in a functional power series in terms of η, $\bar{\eta}$ and I, and then use the definitions of the generating functionals $Z\{\eta, \bar{\eta}, I\}$ and $S\{\eta, \bar{\eta}, I\}$. Under transformation (4. 36) the operators of the functional derivatives are transformed as follows:

$$e^{\bar{\eta} S_F \eta} \frac{\delta}{\delta \bar{\eta}(x)} e^{-\bar{\eta} S_F \eta} = \frac{\delta}{\delta \bar{\eta}(x)} - \int S_F(x-y)\, \eta(y)\, d^4y \equiv i\chi(x), \quad (4.38)$$

$$e^{\bar{\eta} S_F \eta} \frac{\delta}{\delta \eta(x)} e^{-\bar{\eta} S_F \eta} = \frac{\delta}{\delta \eta(x)} + \int \bar{\eta}(y)\, S_F(y-x)\, d^4y \equiv \frac{1}{i} \bar{\chi}(x), \quad (4.39)$$

$$e^{\frac{1}{2} I\Delta_F I} \frac{\delta}{\delta I(x)} e^{-\frac{1}{2} I\Delta_F I} = \frac{\delta}{\delta I(x)} - \int \Delta_F(x-y)\, I(y)\, d^4y \equiv \frac{1}{i} \Phi(x). \quad (4.40)$$

The equations for the functional $S\{\eta, \bar{\eta}, I\}$ are obtained from the equations (4.25), (4.26) and (4.27) for the functional $Z\{\eta, \bar{\eta}, I\}$. Substituting, in equation (4. 25), expression (4. 36) for $Z\{\eta, \bar{\eta}, I\}$, and using formulas (4.38) and (4.40) for the transformation of functional derivatives, we obtain the equation for $S\{\eta, \bar{\eta}, I\}$:

$$\{D(x) - g\gamma_5(x)\, \Phi(x)\}\, \chi(x)\, S\{\eta, \bar{\eta}, I\} = +i\eta(x)\, S\{\eta, \bar{\eta}, I\}. \quad (4.41)$$

The other equations for $S\{\eta, \bar{\eta}, I\}$ are deduced in an analogous fashion

$$\{D(-x) - g\gamma_5(x)\, \Phi(x)\}\, \bar{\chi}(x)\, S\{\eta, \bar{\eta}, I\} = i\bar{\eta}(x)\, S\{\eta, \bar{\eta}, I\}, \quad (4.42)$$

$$\{K(x)\, \Phi(x) + g\, [\bar{\chi}(x)\, \gamma_5 \chi(x)]\}\, S\{\eta, \bar{\eta}, I\} = iI(x)\, S\{\eta, \bar{\eta}, I\}. \quad (4.43)$$

The boundary conditions for the functional $S\{\eta, \bar{\eta}, I\}$ can be obtained from the conditions for $Z\{\eta, \bar{\eta}, I\}$. According to formula (4. 28), for $I=\eta=\bar{\eta}=0$ we should have for the stationary state a:

$$S^{(a)}\{0, 0, 0\}=\delta_{(a)\,(0)}. \tag{4.44a}$$

For the generating functional $S^{(0)}\{\eta, \bar{\eta}, I\}$ of the vacuum functions, to condition (4.29) corresponds the condition

$$\frac{\delta}{\delta\eta}S=0;\ \frac{\delta}{\delta\bar{\eta}}S=0;\ \frac{\delta}{\delta I}S=0 \tag{4.44b}$$

for $\eta=\bar{\eta}=I=0.$

The transition amplitude $U_{aa'}$ can be computed by means of Feynman amplitudes $f_{nml}(x\ldots|y\ldots|z\ldots)$, according to formula (4.12), if we replace in it the function τ_{nml} by the function f_{nml}, since the additional terms which then arise, do not contribute to $U_{aa'}$.

In the limiting case, when all the times $x_0\ldots\ y_0\ldots\ z_0$ are equal and tend to $+\infty$ (the fields ψ, $\bar{\psi}$ and φ are then equal to the free fields ψ_{out}, $\bar{\psi}_{out}$ and φ_{out}) the Feynman amplitude $f_{nml}(x\ldots|y\ldots|z\ldots)$ is, by (4.31), equal to the probability amplitude $\Psi_{nml}(x\ldots|y\ldots|z\ldots)$.

3. The ρ functions /51/. Let us introduce a new functional $R\{\eta, \bar{\eta}, I\}$ by representing the functional Z^0 in the form

$$Z^0\{\eta, \bar{\eta}, I\}=\exp R\{\eta, \bar{\eta}, I\}. \tag{4.45}$$

The functional $R\{\eta, \bar{\eta}, I\}$ will be the generating functional for the functions $\rho_{nml}(x\ldots|y\ldots|z\ldots)$:

$$R=\sum i^{n+m+l}(n!\,m!\,l!)^{-1}\int \rho_{nml}(x\ldots|y\ldots|z\ldots)\times \\ \times\bar{\eta}(x)\ldots\eta(y)\ldots I(z)\ldots d^4x d^4y d^4z. \tag{4.46}$$

The connection between the functions ρ_{nml} and the functions τ_{nml} can be established by expanding the right-hand side of (4.45) in a series and using definitions (4.13) and (4.46). The relations for the first ρ and τ functions are of the form

$$\left.\begin{aligned}
\tau_{200}(x_1x_2|-|-) &= \rho_{100}(x_1|-|-)\,\rho_{100}(x_2|-|-)+\\
&\quad+\rho_{200}(x_1x_2|-|-),\\
\tau_{011}(-|y|z) &= \rho_{011}(-|y|z),\\
\tau_{111}(x|y|z) &= \rho_{100}(x|-|-)\,\rho_{011}(-|y|z)+\rho_{111}(x|y|z),\\
\tau_{211}(x_1x_2|y|z) &= \rho_{100}(x_1|-|-)\,\rho_{100}(x_2|-|-)\,\rho_{011}(-|y|z)+\\
&+\rho_{100}(x_1|-|-)\,\rho_{111}(x_2|y|z)+\rho_{100}(x_2|-|-)\,\rho_{111}(x_1|y|z)+\\
&+\rho_{200}(x_1x_2|-|-)\,\rho_{011}(-|y|z)+\rho_{211}(x_1x_2|y|z)
\end{aligned}\right\} \tag{4.47}$$

From formulas (4.47) it is evident that the ρ functions are equal to zero, if the values of the coordinates x... y... z... allow the representation of the corresponding τ function in the form of a product of τ functions of a smaller number of variables. The τ function decomposes into a product of functions of lower order if the coordinates of a certain group of variables of the τ function are divided by a large space-like or time-like interval (e.g., $|x_1x_2| \gg \left(\frac{h}{mc}\right)^2$. That means that the function ρ will be large only if the intervals between all its coordinates are small. This property of the ρ functions can be useful for the examination of scattering in the case of low energies and large angular momenta. We obtain the equations for the functional $R\{\eta, \bar{\eta}, I\}$, by substituting in (4.25)–(4.27) the expression (4.45) for $Z^0\{\eta, \bar{\eta}, I\}$:

$$\left.\begin{aligned}
\left\{D(x) - ig\gamma_5\frac{\delta}{\delta I(x)}\right\}\frac{\delta}{\delta\bar{\eta}(x)}R &= -\eta(x) + ig\gamma_5\frac{\delta R}{\delta I(x)}\frac{\delta R}{\delta\bar{\eta}(x)},\\
\left\{K(x)\frac{\delta}{\delta I(x)} - ig\frac{\delta}{\delta\eta(x)}\gamma_5\frac{\delta}{\delta\bar{\eta}(x)}\right\}R &= I(x) + ig\frac{\delta R}{\delta\eta(x)}\gamma_5\frac{\delta R}{\delta\bar{\eta}(x)}.
\end{aligned}\right\} \tag{4.48}$$

Thus, equations (4.48) for $R\{\eta, \bar{\eta}, I\}$ are nonlinear. Equations (4.48) serve as a basis for a system of "linked" equations for the ρ functions. In that case, as distinct from the system of equations for the τ functions, the τ function appears only in equations for ρ functions with two coordinates.

§ 5. The space-time treatment of the quantum theory of fields

1. The fundamental equations for the four-vector of state. The space-time description has some essential pecularities as compared with the usual "three-dimensional" description, which may prove important for the further development of the quantum theory of fields. One may assume that the space-time treatment will serve as a formal basis for the concepts of the future field theory. Therefore we shall presently consider in detail the apparatus of such a treatment in connection with the method of functionals /26, 33-35, 37/.

The usual "three-dimensional" formulation of the quantum theory of fields is based on the equations for the field-operators and the canonical commutation relations. The solution of the field equations leads to infinite expressions which are eliminated in the calculation process through the introduction of renormalizing constants. In that case we have to use, in the usual formulation, the notion of masses and charges of "bare" particles, and only at the end of the calculations, after the removal of the divergences, the experimental masses and charges alone are left in the formulas. In other words, by using the equations and the commutation relations for the field operators, it is impossible to avoid divergences and the introduction of the concept of "bare" particles.

The space-time description is not connected with a canonical formalism; in it there are no canonical commutation relations and equations for the field operators. This gives ground to hope that, on the basis of the space-time treatment, we shall succeed in constructing a theory free of divergences and using only the concepts of physical mass and physical charge. This conclusion can be elucidated by means of a visualized representation of an elementary particle as a system of a "bare" particle, surrounded by a "cloud" of other particles. The total mass, charge and spin of this system ought to be equal to the experimental values of the mass, charge and spin of the elementary particle. But one and the same values of these quantities may be obtained for different "clouds" around the "bare" particle. If we consider the evolution in time then, generally speaking (depending on the interaction), the "clouds" around the elementary particle can be different for different moments of time and, as a consequence, different clouds will be assigned at different times to the elementary particle. Therefore, in the three-dimensional treatment, it is impossible to define the "cloud" around the particle for all times. In the four-dimensional treatment however, the evolution in time is not considered and, as a consequence, such a problem does not arise at all. As we shall see below, the functionals in the space-time treatment do not differ from the functionals of the external sources, considered in § 4. The difference between the four-dimensional treatment and the theory with external sources consists in the approach to the description of the field. The theory with external sources is developed on the basis of the usual "three-dimensional" theory /34, 35, 26/. In the space-time treatment, we can introduce from the very beginning the four-vectors of state and its operators. The external sources in it serve for the representation of these operators just as in the method of Fock functionals (§ 2), the quantities $\bar{a}(k)$, $\bar{b}(q)$, $\bar{c}(p)$ serve for the representation of the operators $\psi, \bar{\psi}, \varphi$. From this point of view, the external sources may not be considered as auxiliary quantities in the space-time treatment.

Let us pass to the construction of the apparatus of the space-time treatment in connection with the method of functionals. In the "three-dimensional" treatment, the state vector is defined on a space-like hypersurface;

in the space-time treatment, the state vector Ω has to be defined over the entire four-dimensional volume. This can be done by choosing as the basic field-operators not the ordinary operators $\psi, \bar{\psi}, \varphi$ but other operators $\chi(x), \bar{\chi}(y)$ (the nucleon field) and $\Phi(z)$ (the meson field), which anticommute or commute for any intervals between the points x, y, z*:

$$\left.\begin{array}{ll}\{\chi(x), \bar{\chi}(y)\} = 0, & [\Phi(z), \Phi(z')] = 0, \\ \{\chi(x), \chi(y)\} = 0, & [\chi(x), \Phi(z)] = 0 \text{ и т. д.}\end{array}\right\} \tag{5.1}$$

It has been proposed that the operators χ, $\bar{\chi}$ and Φ be called casual operators /35/.

By virtue of relations (5.1), it is possible to construct, by means of the operators $\chi, \bar{\chi}$ and Φ, a complete system of intercommuting operators $\hat{\xi}$, referring to the four-dimensional volume. We may then take as the basic vectors the eigenvectors $\Omega(\xi)$ of the operators $\hat{\xi}$. The four-state vector

$$\Omega = \int C(\xi)\, \Omega(\xi)\, d\xi \tag{5.2}$$

will be determined if the coefficients of the expansion $C(\xi) = (\Omega(\xi), \Omega)$ are known. The equations for the coefficients of $C(\xi)$ follow from the action principle**.

The action principle in quantum theory has been elaborated in detail by Feynman and Schwinger /27, 28/. If applied to the space-time treatment under consideration, the action principle can be expressed by formula

$$\delta C(\xi) = i\, (\Omega(\xi), \delta W \cdot \Omega), \tag{5.3}$$

where $\delta C(\xi)$ denotes an infinitely small change of $C(\xi)$, caused by the variation of ξ in the four-dimensional volume; W denotes the action operator. Since $\delta C(\xi) = (\delta\Omega(\xi), \Omega)$, it is possible, by introducing the operators of the infinitesimal transformation G_ξ by means of relation

$$(\delta\Omega(\xi), \Omega) = (\Omega(\xi), G_\xi \Omega),$$

----

* The operators χ and $\bar{\chi}$ are not conjugate although, as we shall see below (§ 6), the operators χ and $\bar{\chi}$ can be compared with the operators ψ and $\bar{\psi}$ of the three-dimensional treatment.

** An attempt to obtain the equations, not relying on the Lagrangian formalism, has been made in reference /32/.

to represent the action principle (5.3) in the form

$$\delta\Omega = G_\xi \Omega = i\delta W \cdot \Omega. \tag{5.4}$$

The variation $\delta\Omega$ in (5.4) is derived from the variation of the coefficients $C(\xi)$.

We shall assume that the action W has the same form as in the usual "three-dimensional" treatment, but that it is composed of the operators $\chi, \bar{\chi}$ and Φ. Equation (5.4) can readily then be solved symbolically in a representation in which the operators χ, χ and Φ are operators of multiplication by the quantities $\chi', \bar{\chi}'$ and by the function $\bar{\Phi}'$, and the four-state vector Ω is a functional of $\bar{\chi}', \chi'$ and Φ'. In this representation the operator W will consist only of operators of multiplication and, as a consequence, equation (5.4) has the solution

$$\Omega\{\chi', \bar{\chi}', \Phi'\} = e^{iW'} \cdot \frac{1}{N}, \tag{5.5}$$

where W' consists of $\chi', \bar{\chi}'$ and Φ', and the constant N^{-1} is Ω for $\chi' = \bar{\chi}' = \Phi' = 0$.

It is convenient to represent the action W in the form

$$W = \int L(x, y)\, d^4x d^4y,$$

$$L(x, y) = \frac{1}{2} i\delta^4(x - y) \{\bar{\chi}(y)[D(x) - g\Phi(x)\gamma_5]\chi(x) - \chi(y)[D(-x) - g\Phi(x)\gamma_5(x)]\bar{\chi}(x) - \Phi(x)K(y)\Phi(y)\}. \tag{5.6}$$

From (5.5) we can obtain symbolic solutions for other representations.

If the operators $\hat{\xi}$ are constructed from the operators $\chi, \bar{\chi}$ and Φ we can express the variation $\delta\hat{\xi}$ formally in terms of $\delta\chi$, $\delta\bar{\chi}$ and $\delta\Phi$ (the meaning of the variation of the Fermi-field operators χ and $\bar{\chi}$ shall be considered in § 6). The variations $\delta\chi$, $\delta\bar{\chi}$ and $\delta\Phi$ satisfy the same commutation relations (5.1) as the operators $\chi, \bar{\chi}, \Phi$. We introduce, for the construction of the operator of infinitesimal transformation G_ξ, the operators $\pi, \bar{\pi}$ and Π, which we define by means of the commutation relations:

$$\left.\begin{aligned} \{\pi(x), \chi(y)\} &= -i\delta^4(x - y), \quad [\Pi(x), \Phi(y)] = -i\delta^4(x - y), \\ \{\bar{\pi}(x), \bar{\chi}(y)\} &= i\delta^4(x - y). \end{aligned}\right\} \tag{5.7}$$

The operators π, $\bar{\pi}$ and Π are the four-dimensional analogues of the canonical conjugate momenta of the "three-dimensional" theory. In the representation with the functional $\Omega\{\chi', \bar{\chi}', \Phi'\}$ (formula (5.5), π, $\bar{\pi}$ and Π are the operators of the functional derivatives. Fundamental equation (5.4) may then be written in the form

$$[G_{\chi} + G_{\bar{\chi}} + G_{\Phi}]\,\Omega = -i\,(\delta_{\chi}W + \delta_{\bar{\chi}}W + \delta_{\Phi}W)\,\Omega, \tag{5.8}$$

where the operators G_{χ}, $G_{\bar{\chi}}$ and G_{Φ} have the properties

$$[G_{\chi}, \chi] = \delta\chi; \quad [G_{\bar{\chi}}, \bar{\chi}] = \delta\bar{\chi}; \quad [G_{\Phi}, \Phi] = \delta\Phi,$$

and $\delta_{\chi}W = [G_{\chi}, W]$ etc. The operators of the infinitesimal transformation G_{χ}, $G_{\bar{\chi}}$ and G_{Φ} have the form

$$\left.\begin{aligned} G_{\chi} &= i\int \delta\chi(x)\,\pi(x)\,d^4x, \\ G_{\bar{\chi}} &= -i\int \delta\bar{\chi}(x)\,\bar{\pi}(x)\,d^4x, \end{aligned}\quad G_{\Phi} = i\int \delta\Phi(x)\,\Pi(x)\,d^4x. \right\} \tag{5.9}$$

As a consequence of the independence of the variations $\delta\chi$, $\delta\bar{\chi}$ and $\delta\bar{\Phi}$ we obtain from formula (5.8) three equations

$$\left.\begin{aligned} \{D(x) - g\gamma_5\Phi(x)\}\,\chi(x)\,\Omega &= i\bar{\pi}(x)\,\Omega; \\ \{D(-x) - g\gamma_5(x)\,\Phi(x)\}\,\bar{\chi}(x)\,\Omega &= i\pi(x)\,\Omega; \\ \{K(x)\,\Phi(x) + g\bar{\chi}(x)\,\gamma_5\chi(x)\}\,\Omega &= i\Pi(x)\,\Omega. \end{aligned}\right\} \tag{5.10}$$

Let us note that equations (5.10) have been obtained here without making recourse to the usual "three-dimensional" apparatus of field theory. Let us now consider a representation where the operators π, $\bar{\pi}$ and Π are multiplication operators, χ, $\bar{\chi}$ and Φ being, in view of (5.6) the operators of functional derivatives. Hence, comparing (5.10) with (4.25)-(4.27), it can readily be established that equations (5.10) represent nothing else but the equations for the functional of external sources $Z\{\eta, \bar{\eta}, I\}$.

Thus we have found the correspondence between the space-time and the ordinary three-dimensional treatments: the four-state vector Ω in a representation where the application of the operators π, $\bar{\pi}$ and Π consists in multiplication by the functions of external sources (here $\bar{\eta} = \eta^*\gamma_4$):

$$\pi(x)\,\Omega = \eta(x)\,\Omega; \quad \bar{\pi}(x)\,\Omega = \eta(x)\,\Omega; \quad \Pi(x)\,\Omega = I(x)\,\Omega, \tag{5.11}$$

is proportional to the Schwinger functional of external sources $Z\{\eta; \bar{\eta}, I\}$.

The causes of such a correspondence can be understood by considering once again the action principle (5.4) or (5.3). In this form, the action principle presupposes that the variation of the field quantities (e.g., varieties $\delta\chi$, $\delta\bar{\chi}$ and $\delta\Phi$) at any point of the four-dimensional volume can affect the state vector Ω. In other words, the principle of stationary action which leads in the usual "three-dimensional" treatment to the equations for the field operators, is denied in the space-time treatment. Hence follows, in particular, the absence of equations of motion for the operators χ, $\bar{\chi}$ and Φ (there exists only a condition for determining possible functionals, i.e., equation (5.4)). Since the principle of stationary action is denied in the four-dimensional treatment, i.e., variations of the field quantities are allowed which disturb the principle of stationary action, the apparatus of the four-dimensional treatment is equivalent to such an apparatus of the ordinary "three-dimensional" theory where variations are considered which likewise disturb the principle of stationary action; of this kind are the variations of the field quantities, caused by the variations of the external parameters. A "three-dimensional" treatment with external sources corresponds therefore to the space-time treatment.

2. <u>The generalized Fock functional</u>. The correspondence between the four-dimensional formalism and the apparatus of the ordinary "three-dimensional" theory can be represented in a more convenient and obvious manner, by passing to another representation in which the four-dimensional state vector will be the generalized Fock functional. We recall that the Fock functional is the generating functional for probability amplitudes and, at the same time, the state vector in a representation in which the creation operators $a^+(p)$, $b^+(q)$ and $c^+(k)$ are multiplication operators. For the purpose of constructing the generalized Fock functional in the space-time treatment, we introduce the "four-dimensional" creation operators $a_\rho^+(x)$, $b_\lambda^+(y)$ (for the nucleon field) and $\overset{+}{c}(z)$ (for the meson field) (λ and ρ are spinor indexes). Unlike the ordinary commutation relations, the commutation between the "four-dimensional" creation operators and the hermitian-conjugate annihilation operators $a_{\rho'}(x)$, $b_{\lambda'}(y)$, $c(z)$ contains the δ function in its right-hand side:

$$\left.\begin{array}{ll}\{a_\rho(x),\ a_\sigma^+(y)\} = \delta_{\rho\sigma}\delta^4(x-y), & \\ \{b_\rho(x),\ b_\lambda^+(y)\} = \delta_{\rho\lambda}\delta^4(x-y), & [c(z),\ c^+(z')] = \delta^4(z-z').\end{array}\right\} \qquad (5.12)$$

Proceeding from the definition of the "four-dimensional" creation and annihilation operators (5.12), we can formally carry over into the space-time theory all the mathematical results of the method of Fock functionals (§2). We choose a representation in which $\overset{+}{a}(x)$, $\overset{+}{b}(y)$, $\overset{+}{c}(z)$ are operators of multiplication by the anticommuting quantities $\bar{a}(x)$, $\bar{b}(y)$ and by the function

$\bar{c}(z)$. The four-state vector $F\{\bar{a}, \bar{b}, \bar{c}\}$, which is a functional of $\bar{a}$, $\bar{b}$, $\bar{c}$ may then be represented in the form of an expansion in terms of the eigenfunctionals of the number operators of nucleons $\int a^+(x)a(x)d^4x$, and of antinucleons $\int b^+(x)b(x)d^4x$ and of mesons $\int c^+(x)c(x)d^4x$:

$$F=\sum F_{nml}=\sum_{nml}(n!\,m!\,l!)^{-1}\int \bar{a}(x_n)\ldots f_{nml}(x_1\ldots x_n|\,y_1\ldots y_m|\,z_1\ldots z_l)\times \times\bar{b}(y_m)\ldots\bar{c}(z_1)\ldots d^4x_1\ldots d^4y_1\ldots d^4z_1\ldots \tag{5.13}$$

The operators a, b, c are, in relation to the functional (5.13), the operators of functional derivatives:

$$a(x)=\frac{\delta}{\delta\bar{a}(x)};\quad b(y)=\frac{\delta}{\delta\bar{b}(y)};\quad c(z)=\frac{\delta}{\delta\bar{c}(z)}. \tag{5.14}$$

The functions f_{nml}, for which the generalized Fock-functional is a generating functional, can be considered as the four-dimensional analogues of the probability amplitues. As we shall see below, f_{nml} is a Feynman amplitude. The functional (5.13) was introduced by Coester /37/.

Let us consider the energy-momentum vector. In the ordinary "three-dimensional" treatment, the energy-momentum vector can be found if the Lagrangian function is known and if the properties of the energy-momentum vector as a displacement vector are the consequences of the commutation relations. In the "four-dimensional" treatment of the field theory there is no variation principle and, as a consequence, there are no canonical commutation relations and equations for the field operators. We define therefore the energy-momentum vector $P_\mu\,(iP_0=P_4)$ as a quantity which has the properties of the displacement vector:

$$-i\,[P_\mu,\,a(x)]=\frac{\partial a(x)}{\partial x_\mu}\ \text{etc} \tag{5.15}$$

with intercommuting components: $[P_\mu,\,P_\nu]=0$.

The expression for P_μ has the form

$$P_\mu=-i\int\left\{a^+(x)\frac{\partial a(x)}{\partial x_\mu}+b^+(x)\frac{\partial b(x)}{\partial x_\mu}+c^+(x)\frac{\partial c(x)}{\partial x_\mu}\right\}d^4x. \tag{5.16}$$

The application of the operator P_μ to the state vector (5.13) is equivalent to the differentiation of the amplitudes $f(x\ldots|y\ldots|z\ldots)$: if $P_\mu F = F'$ and $f'_{nml}(x\ldots|y\ldots|z\ldots)$ are amplitudes of the functional F', then

$$f'_{nml}(x\ldots|y\ldots|z\ldots)=$$
$$=-i\sum\left(\frac{\partial}{\partial x_\mu}+\ldots+\frac{\partial}{\partial y_\mu}+\ldots+\frac{\partial}{\partial z_\mu}+\ldots\right)f_{nml}(x\ldots|y\ldots|z\ldots). \quad (5.17)$$

Let us define the vector of the "vacuum" state F_0. By close analogy with the three-dimensional treatment, we put

$$a(x)F_0=0;\quad b(x)F_0=0;\quad c(z)F_0=0;\quad (F_0,\ F_0)=1. \quad (5.18)$$

It is obvious that $P_\mu F_0 = 0$, where the state F_0 will have the least energy if all the amplitudes $f_{nml}(x\ldots|y\ldots|z\ldots)$ contain only positive frequencies. Hitherto we used for the construction of the generalized Fock functional only the commutation relations (5.24). In order to establish the correspondence with the functional Ω (see § 5, 1), we have to express the causal operators χ, χ and Φ in terms of the creation operators $\overset{+}{a}$, $\overset{+}{b}$, $\overset{+}{c}$ and annihilation operators a, b, c. For this purpose it is necessary to assume that, in a representation with (5. 13), it is possible to subdivide the field operators χ, $\bar{\chi}$ and Φ into creation and annihilation parts, where all the creation parts and all the annihilation parts commute or anticommute separately. Formulas (5. 1) and (5. 12) will hold only if the commutations between the creation parts χ^c, $\bar{\chi}^c$, Φ^c and the annihilation parts χ^a, $\bar{\chi}^a$, Φ^a are equal to certain functions σ_F and d_F:

$$\left.\begin{array}{ll}\{\chi^a(x),\ \bar{\chi}^c(y)\}=\sigma_F(x,\ y); & \\ \{\bar{\chi}^a(x),\ \chi^c(y)\}=-\sigma_F(y,\ x); & [\Phi^a(x),\ \Phi^c(y)]=d_F(x,\ y)=d_F(y,\ x).\end{array}\right\} \quad (5.19)$$

This means that, in relation to the generalized Fock-functional F, the operators χ, $\bar{\chi}$ and Φ can be in the form

$$\left.\begin{array}{l}\chi(x)=a(x)-\int\sigma_F(x,\ y)\,b^+(y)\,d^4y,\\ \bar{\chi}(x)=b(x)+\int a^+(y)\,\sigma_F(y,\ x)\,d^4y,\\ \Phi(x)=c(x)+\int d_F(x,\ y)\,c^+(y)\,d^4y.\end{array}\right\} \quad (5.20)$$

It is now possible to establish the connection between the generalized Fock functional F and the functional of external sources Ω (see formulas (5.10) and (5.11)), in relation to which the operators χ, $\bar{\chi}$ and Φ are operators of functional. The functionals F and Ω are connected by the transformation

$$F=R\Omega, \quad (5.21a)$$

where

$$R = \exp\left\{\int\left[a^{+}(x)\,\sigma_F(x, y)\,b^{+}(y) - \right.\right.$$
$$\left.\left. - \frac{1}{2}\,c^{+}(x)\,d_F(x, y)\,c^{+}(y)\right]d^4x d^4y\right\}, \qquad (5.21b)$$

whence $\pi = -ia^{+}$; $\bar{\pi} = ib^{+}$; $\Pi = ic^{+}$.

Transformation (5.21) corresponds to transformation (4.36) of the generating functional for T-functions to the generating functional for Feynman amplitudes, if σ_F and d_F denote the Feynman-Green's functions S_F and Δ_F which, moreover, follows from (5.19) too.

The correspondence between the four-dimensional formalism and the apparatus of the ordinary theory can now be expressed by the equation /37/:

$$\begin{aligned} T_{nml}(x_1 \ldots x_n \,|\, y_1 \ldots y_m \,|\, z_1 \ldots z_l) &\equiv \\ \equiv (\Psi_0, T\,[\psi(x_1) \ldots \psi(x_n)\,\bar{\psi}(y_1) \ldots \bar{\psi}(y_m)\,\varphi(z_1) \ldots \varphi(z_l)]\,\Psi) &= \qquad (5.22) \\ = (F_0, \chi(x_1) \ \ldots \ \chi(x_n)\,\bar{\chi}(y_1) \ \ldots \ \chi(y_m)\,\Phi(z_1) \ldots, \ \Phi(z_l)\,F\,\{\bar{a}, \bar{b}, c\}), & \end{aligned}$$

where Ψ_0, Ψ are the state vectors, and ψ, $\bar{\psi}$ and φ the field operators in the Heisenberg representation; in the right-hand side we have the matrix element of the "four-dimensional" theory. If we use transformation (5.21), we find another formula which connects the matrix elements of the four-dimensional and three-dimensional treatments:

$$\begin{aligned} (\Psi_0, T\,[\psi(x_1) \ \ldots \ \psi(x_n)\,\bar{\psi}(y_1) \ \ldots \ \bar{\psi}(y_m)\,\varphi(z_1) \ \ldots \ \varphi(z_l)]\,\Psi) &= \\ = (F_0, \chi(x_1) \ldots \chi(x_n)\,\bar{\chi}(y_1) \ldots \bar{\chi}(y_m)\,\Phi(z_1) \ \ldots \ \Phi(z_l)\,\Omega\,\{\eta, \bar{\eta}, I\}). & \qquad (5.23) \end{aligned}$$

In (5.22), the operators χ, $\bar{\chi}$, Φ ought to be represented in the form (5.20); in (5.23), the operation of χ, $\bar{\chi}$ and Φ on Ω and F_0 is determined by formulas (5.7), (5.11) and (5.21b).

3. The functional Fourier transformation. In § 5, 1 we obtained expression (5.5) for the state vector $\Omega\{\chi', \bar{\chi}'\Phi'\}$ for the case in which operators χ, $\bar{\chi}$ and Φ are functions, and π, $\bar{\pi}$ and Π are operators of functional. However, to the real case corresponds, as has been established in the same paragraph, a representation in which χ, $\bar{\chi}$ and Φ are not functions but operators of functional, the operators of multiplication being π, $\bar{\pi}$ and Π. The general solution of the problem of interacting fields, i.e., the solution of equation (5.10) under condition (5.11), can therefore be obtained by means of a functional-Fourier transformation from the previously found functional (5.5).

Let $F\{I\}$ be the functional of the function $I(x)$, depending on the space-time variable x. Let us divide the entire space-time manifold into n cells of equal volume and let us consider, instead of the functions $I(x)$, the set of their mean values $I_1, \ldots I_k, \ldots I_n$ in the respective cells (k is the number of the cell). The functional $F\{I\}$ will then be a function $F(I_1 \ldots I_n)$ of the values $I_1 \ldots I_n$. The Fourier-transformation for $F(I_1 \ldots I_n)$ has the form

$$F(I_1 \ldots I_n) = \int e^{-i \Sigma I_k \Phi'_k} F(\Phi'_1 \ldots \Phi'_n) \frac{d\Phi'_1}{\sqrt{2\pi}} \cdots \frac{d\Phi'_n}{\sqrt{2\pi}},$$

where Φ'_k also refers to the k-th cell. In passing to the limit, when each cell contains only one space-time point, we obtain for $F\{I\}$ a representation by means of the integral

$$F\{I\} = \int e^{-i \int I(x) \Phi'(x) d^4 x} F\{\Phi'\} d(\Phi'), \tag{5.24}$$

where in $d(\Phi') = \prod_k \frac{d\Phi'_k}{\sqrt{2\pi}}$ the index k runs over all space-time points. The application of the operator

$$\Phi(x) = i \frac{\delta}{\delta I(x)}$$

to the functional F leads to the appearance of a factor Φ' in the integrand. We can say that $F\{\Phi'\}$ is a functional in a representation in which the multiplication operator is $\Phi(x)$, and $F\{I\}$ is a functional in a representation in which Φ is the operator of the functional derivative with respect to I.

By analogy, the Fourier transform of the functional $\Omega\{\eta, \bar{\eta}, I\}$ is the functional $\Omega\{\chi', \bar{\chi}', \Phi'\}$, which is the four-state vector in a representation in which $\chi, \bar{\chi}$ and Φ are multiplication operators. Therefore the general solution of equation (5.10) for the functional $\Omega\{\eta, \bar{\eta}, I\}$ can be represented in the form of the following integral:

$$\Omega\{\eta, \bar{\eta}, I\} = \int e^{-i \int (\eta \bar{\chi}' + \chi \bar{\eta} + \Phi' I) d^4 x} \Omega\{\chi', \bar{\chi}', \Phi'\} d(\chi') d(\bar{\chi}') d(\Phi'). \tag{5.25}$$

Substituting for $\Omega\{\chi', \bar{\chi}', \Phi'\}$ the expression (5,5), we find that

$$\Omega\{\eta, \bar{\eta}, I\} = \frac{1}{N} \int e^{-i \int (\eta \bar{\chi}' + \chi \bar{\eta} + \Phi' I) d^4 x} e^{iW'} d(\chi') d(\chi') d(\Phi'). \tag{5.26}$$

where the constant N * equal to $\Omega\{\chi', \bar{\chi}', \Phi'\}$ for

$$\chi' = \bar{\chi}' = \Phi' = 0$$

(see (5.5), is determined by the normalizing condition.

We are interested in the generating functional for the vacuum T-functions. If we denote the corresponding four- state vector by Ω^0, we obtain from the correspondence formulas (5.23) the normalizing condition for Ω^0:

$$(F_0, \Omega^0\{\eta, \bar{\eta}, I\}) = 1. \qquad (5.27a)$$

In view of definition (5.18) of the vector of the "vacuum" state F_0, which is, by (5.21b) and (5.11) equivalent to the equations $(F_0, \pi(x)\Omega') = (F_0, \bar{\eta}(x)\Omega') = 0$ and so on, (for an arbitrary functional Ω'), we can write the normalizing condition (5.29a) in the same form as the boundary conditions for the function $Z\{\eta, \bar{\eta}, I\}$ (see § 4, 1):

$$\Omega^0\{0, 0, 0\} = 1. \qquad (5.27b)$$

Hence the functional Ω^0 coincides with Z.

From condition (5.27b) we obtain the normalizing constant:

$$N = \int e^{iW} d(\chi') d(\bar{\chi}') d(\Phi'). \qquad (5.28)$$

We hence obtain for the T-functions, from (5.23) and (5.25) an expression in the form of a integral:

$$T(x\ldots|y\ldots|z\ldots) = (F_0, \chi(x)\ldots\bar{\chi}(y)\ldots\Phi(z)\ldots\Omega\{\eta, \bar{\eta}, I\}) =$$
$$= \frac{1}{N}\int \chi'(x)\ldots\bar{\chi}'(y)\ldots\Phi'(z)\ldots e^{iW} d(\chi') d(\bar{\chi}') d(\Phi'), \qquad (5.29)$$

which can be interpreted as the product $[\chi'(x)\ldots\bar{\chi}'(y)\ldots\Phi'(z)\ldots]$ averaged over the Fermi and Bose fields; in this case the action-exponent plays the role of the weight function.

In formulas of type (5.28) and (5.29) we can readily perform either integration over a Fermi field or integration over a Bose field. If the T-function does not depend on meson coordinates, then, by introducing a new variable Φ_1 instead of Φ' according to formula

$$\Phi'(x) = \Phi_1(x) + \frac{g}{2}\int \Delta_F(x-y)\bar{\chi}(y)\gamma_5\chi(y)d^4y, \qquad (5.30)$$

we obtain:

$$T(x\ldots|y|\ldots|-)=\frac{1}{N}\int\chi'(x)\ldots\bar{\chi}'(y)\ldots e^{iW_1}d(\chi')\,d(\bar{\chi}'), \tag{5.31}$$

where

$$W_1=i\int\bar{\chi}(x)\,D(x)\,\chi(x)\,d^4x-\frac{ig^2}{2}\int\bar{\chi}(x)\,\gamma_5\,\chi(x)\times$$
$$\times\Delta_F(x-y)\,\bar{\chi}(y)\,\gamma_5\,\chi(y)\,d^4x\,d^4y.$$

The derivation of formulas (5.29) and (5.33) was not based on perturbation theory and therefore, their application does not depend on the value of the coupling constant. In that consists the value of formulas of analogous type. Similar formulas can be used as points of departure in attempts of developing an approximation method which is different from the usual perturbation method*.

We should however, in our case, take into consideration that the integrals χ' and $\bar{\chi}'$ over the Fermi field, are of a symbolic character, since χ' and $\bar{\chi}'$ are anticommuting functions. Therefore, the problem of integration over a Fermi-field requires a special investigation, which will be conducted in § 6.

§ 6. The variation of the operator and functional integration in the case of a Fermi field

The symbolic character of integration over a Fermi field in the solution (5.25), is closely connected with the symbolic character of equations (5.10) which contain variation derivatives with respect to anticommuting functions. We shall see below how it is possible to reduce integration over a Fermi field to ordinary functional integration and how to write equations (5.10) without derivatives with respect to anticommuting functions. For the elucidation of these problems we have to consider the variation of the Fermi field operator.

In the case of a Bose field, the determination of the variation of the field operator does not cause difficulties since there exists a representation in which the field operator is the operator of multiplication by some (auxiliary) function. In the case of a Bose field it is possible to represent the variation of the field $\delta\Phi'$ in, e.g., the following manner if we expand the Bose field $\Phi'(x)$ in a series in terms of the system of basic functions φ_n /53/

----

* It may be that in the modern quantum theory of fields there are no domains of solution for the meson field at all /30/.

$$\Phi'(x) = \sum_n \gamma_n \varphi_n(x),$$

then

$$\delta\Phi'(x) = \sum_n \delta\gamma_n \varphi_n(x),$$

and integration over $\Phi'(x)$ will thus be replaced by integration over γ_n.

However difficulties arise in the case of a Fermi field, since the variation of the operator has to commute with the operator itself; therefore we can proceed in a similar manner only if it is possible to define a complete system of anticommuting (basis) functions $\chi_n(x)$, $\bar{\chi}_n(x)$. We shall then have for the arbitrary Fermi field χ and $\bar{\chi}$ the following expansion

$$\chi(x) = \sum_n \alpha_n \chi_n(x); \qquad \bar{\chi}(x) = \sum_n \beta_n \chi_n(x) \tag{6.1}$$

($\bar{\chi}_n(x)$ is, in general, not conjugate to $\chi_n(x)$; (see footnote to formula (5.1)). If representation (6.1) for χ and $\bar{\chi}$ is possible, the variations $\delta\chi$ and $\delta\chi$ will be expressed in terms of the variations of the numbers α_n and β_n:

$$\delta\chi(x) = \sum \delta\alpha_n \chi_n(x); \qquad \delta\bar{\chi}(x) = \sum \delta\beta_n \bar{\chi}_n(x), \tag{6.2}$$

and functional integration over the anticommuting quantities $\chi(x)$ and $\bar{\chi}(x)$ reduces to integration over the numbers α_n and β_n. Thus the problem reduces to the determination of the basic anti-commuting functions $\chi_n(x)$ and $\bar{\chi}_n(x)$.

Let us introduce a complete system of orthogonal functions $\Psi_n(x)$, which are normalized by the condition

$$\int \bar{\Psi}_n(x) Z(x, y) \Psi_m(y) d^4x d^4y = \delta_{nm} \tag{6.3}$$

(here $\bar{\Psi}_n = \Psi_n^+ \gamma_4$) We take as Z (x, y) either the Dirac operator $-i\delta^4(x-y)D(y)$, or the Dirac operator with external meson field

$$D(x, y, \Phi) = -i\delta^4(x-y)[D(y) - g\gamma_5\Phi(y)],$$

since we are not interested in the case when the action on the nucleon field is different from zero. In addition, we introduce the anticommuting operators A_n and B_n

$$A_n = a_n + b_n^+; \quad B_m = b_m - a_m^+, \tag{6.4}$$

where a_n^+, B_n^+ are "creation" operators, and a_n and b_n are the ("four-dimensional") annihilation operators:

$$\{a_n, a_m^+\} = \{b_n, b_m^+\} = \delta_{nm},$$

so that

$$a_n F_0 = b_n F_0 = 0$$

(F_0 is the "vacuum functional", see (5.18)). We form the operators

$$\chi_n(x) = A_n \Psi_n(x); \qquad \bar{\chi}_n(x) = B_n \bar{\Psi}_n(x), \tag{6.5}$$

which, as we shall see below, can be used as basic anticommuting functions. In order that χ_n and $\bar{\chi}_n$ could be considered as basic functions and that expansion (6.1) should hold, the quantities χ_n and $\bar{\chi}_n$ (or A_n and B_n) ought to be connected by a normalizing condition, which is possible only if we can limit ourselves to the consideration of some class of matrix elements of the operators χ_n and $\bar{\chi}_n$.

In order to elucidate the problem of normalization for χ_n and $\bar{\chi}_n$ (or for A_n and B_n) we shall show that an arbitrary functional F can be expressed in terms of the functional F^0, which is determined by the equations of motion (taking interaction into account) and by the conditions $(F_0, F^0) = 1$, $F_0, a(x)F^0) = (F_0, b(x)F^0) = (F_0, c(x)F^0) = 0$. Equations (5.10) have a formal solution of the form

$$\Omega(g) = S\Omega(0),$$

where the functional for non-interacting fields $\Omega(0)$ satsifies (5.10) for $g = 0$, and the operator S is the analogue of the scattering matrix

$$S = S_0 \exp\left\{g \int N[\bar{\chi}(x)\gamma_5\chi(x)]\,\Phi(x)\,d^4x\right\}, \tag{6.6}$$

where S_0 is a normalizing constant. Since F(g) is connected with $\Omega(g)$ by transformation (5.21), $F = R\Omega$, the generalized Fock functional F(g) is expressed in terms of the functional F(0), in the case of no interaction, by formula /37, 35, 38/.

$$F(g) = RSR^{-1}F(0) \equiv S'F(0). \tag{6.7}$$

whereas from (5.21) and (5.10) it follows that the equations for F(0) have the form

$$\left.\begin{aligned} D(x)\chi(x)F(0) &= -b^{+}(x)F(0), \\ D(-x)\bar{\chi}(x)F(0) &= a^{+}(x)F(0), \\ K(x)\Phi(x)F(0) &= -c^{+}(x)F(0), \end{aligned}\right\} \quad (6.8)$$

i.e., the amplitudes $f^0(x\ldots|y\ldots|z\ldots)$ of F(0) satsify the equations for free fields. In particular, the generating functional $F^0(g)$ for vacuum Feynman amplitudes is

$$F^0(g) = S' F^0(0) = S' F_0. \quad (6.9)$$

Let us now consider as an example $F(g) = S'F_{110}(0)$, where

$$F_{110}(0) = \int f_0(x|y|-)a^{+}(x)b^{+}(y)d^4xd^4y\,F_0.$$

It follows then from (6.8), (6.9) and the commutativity of S' with $\bar{\chi}$ and χ that

$$F(g) = S' F_{110}(0) =$$
$$= -\int f^0(x|y|-)\,D(-x)\bar{\chi}(x) = D(y)\chi(y)d^4xd^4y\,F^0(g). \quad (6.10)$$

Generalization of (6.10) presents no difficulties. Formulas of type (6.10) reduce the problem of matrix elements $(F_0, \chi(x)\ldots\chi(y)\ldots\Phi(z)\ldots F(g))$ for an arbitrary F(g), to the problem of matrix elements between the states F_0 and $F^0(g)$. Since $F^0(g)$ is the result of the application of the operator S' which depends on χ, $\bar{\chi}$, Φ to the "vacuum" functional $F_0 = F^0(0)$ then, as a consequence, all the different matrix elements of χ_n, $\bar{\chi}_m$, which appear in the calculation of the I-functions or of the functionals F and Ω can be reduced to the vacuum of the products of the operators χ_n and $\bar{\chi}_m$:

$$\langle \bar{\chi}_{m_1}(y_1)\bar{\chi}_{m_2}(y_2)\ldots\chi_{n_2}(x_2)\chi_{n_1}(x_1)\rangle_0,$$

or

$$\langle B_{m_1} B_{m_2} \ldots A_{n_2} A_{n_1}\rangle_0,$$

if we denote

$$\langle M\rangle_0 = (F_0, MF_0).$$

The vacuum average of the products of the operators B_m and A_n can be represented in the form of a determinant, composed of $\langle B_m A_n\rangle_0$:

$$\langle B_{m_1} \ldots B_{m_r} A_{n_r} \ldots A_{n_1} \rangle_0 = \begin{vmatrix} \langle B_{m_1} A_{n_1} \rangle_0 \langle B_{m_1} A_{n_2} \rangle_0 \ldots \langle B_{m_1} A_{n_r} \rangle_0 \\ \langle B_{m_2} A_{n_1} \rangle_0 \langle B_{m_2} A_{n_2} \rangle_0 \ldots \langle B_{m_2} A_{n_r} \rangle_0 \\ \ldots \ldots \ldots \ldots \ldots \\ \langle B_{m_r} A_{n_1} \rangle_0 \langle B_{m_r} A_{n_2} \rangle_0 \ldots \langle B_{m_r} A_{n_r} \rangle_0 \end{vmatrix} \tag{6.11}$$

Since $\langle B_m A_n \rangle_0 = \delta_{nm}$, in the calculation of the vacuum average the operator $B_n A_n$ can be replaced by the unit operator, the result (6.11) can also be obtained by means of the rule that in the vacuum average of the products of operators $B_{m1} Bm_2 \ldots A_{n1} A_{n2} \ldots$

$$B_n A_n = -A_n B_n \to 1 \tag{6.12}$$

the product of two anticommuting operators B_n and A_n with similar indexes n is equivalent to the unit operator.

Formula (6.12) can be called the normalizing condition for the operators A_n and B_n, since (6.3) can now be written in the form

$$\int \bar{\chi}_n(x) Z(x, y) \chi_m(y) d^4 x d^4 y \to \delta_{nm}. \tag{6.13}$$

Normalizing condition (6.13) has been proposed by Matthews and Salam /36/. This condition too should be considered not literally, but in the same sense as (6.12). For the deduction of equations of type (5.10), we now need not introduce anticommuting functional derivatives. Let us form the operators of infinitesimal transformations G_χ and $G_{\bar{\chi}}$ (see § 5 1, formula (5.9)).

If we consider only functions of χ, $\bar{\chi}$ and Φ then G_χ and $G_{\bar{\chi}}$ can be represented in the form

$$\left. \begin{aligned} G_\chi &= \sum \delta\alpha_n \frac{\partial}{\partial \alpha_n}; \\ G_{\bar{\chi}} &= \sum \delta\beta_n \frac{\partial}{\partial \beta_n}, \end{aligned} \right\} \tag{6.14}$$

where only derivatives with respect to numbers are included. From the action principle (5.8) we hence obtain, instead of the first two equations (5.10), the equations with ordinary derivatives:

$$\left. \begin{aligned} i \frac{\partial}{\partial \alpha_n} \Omega &= \sum_m \beta_m Q_{mn} \Omega, \\ i \frac{\partial}{\partial \beta_n} \Omega &= \sum_m \alpha_m Q_{nm} \Omega, \end{aligned} \right\} \tag{6.15}$$

where

$$Q_{nm} = \int \bar{\chi}_n(x) D(x, y, \Phi') \chi_m(y) d^4 x d^4 y. \tag{6.16}$$

The solution of equations (6.15) and of the third (meson) equation (5.10) is given by the functional

$$\Omega\{\alpha, \beta, \Phi'\} = \frac{1}{N} \exp\left\{-i \sum_{nm} \beta_m Q_{mn} \alpha_n + iW_M\right\}. \tag{6.17}$$

where W_M is the action for the meson field; N is the normalizing constant. The functional $\Omega\{\alpha, \beta, \Phi'\}$ corresponds to the state vector $\Omega\{\chi', \bar{\chi}', \Phi'\}$ (formula (5.5)) in a representation where $\chi, \bar{\chi}, \Phi$ are multiplication operators, since α and β are considered to be numbers. The formal solution of the problem of interacting fields was obtained in § 5, 3 by means of a Fourier transformation for the functional $\Omega\{\chi', \bar{\chi}', \Phi'\}$, depending on the anticommuting functions $\chi'(x)$, $\bar{\chi}'(x)$ and on the function I(x). For the solution of this problem, it is now necessary to perform the Fourier transformation for the functional $\Omega\{\alpha, \beta, \Phi'\}$ depending on the numbers α_n, β_n and on the function $\Phi'(x)$. Since the quantity Q_{nm} can be considered, by (6.13), as real, the Fourier transformation for $\Omega\{\alpha, \beta, I\}$ will be possible if

$$\alpha_n = \beta_n^* \tag{6.18}$$

and it is assumed that in the operators D(x) and K(x) the masses m and μ^2 are replaced by $m - i\varepsilon$ and $\mu^2 - i\varepsilon$, , since only then the functional integrals over α, β $I(x)$ will converge.

As a result we now obtain $\Omega\{\eta, \bar{\eta}, I\}$ instead of the anticommuting functions $\eta(x)$ and $\bar{\eta}(x)$, a functional which depends on the functions I(x) and the numbers ζ_n, ζ_n^*:

$$\omega\{\zeta^*, \zeta, I\} = \int e^{-i\sum\left(\zeta_n^* \beta_n^* + \zeta_n \beta_n\right) - i\int I(x)\Phi(x)d^4x}\, \Omega\{\beta^*, \beta, \Phi\}\, d(\beta)\, d(\Phi), \tag{6.19}$$

$$\Omega\{\beta^*, \beta, \Phi\} = \frac{1}{N} \exp\left[-i \sum_{nm} Q_{nm} \beta_n \beta_m^* + iW_M\right]. \tag{6.20}$$

which differs from (5.25) by the fact that in (6.19) we integrate over the complex numbers β_n:

$$d(\beta) = \prod_n \frac{d\beta_n}{\sqrt{2\pi}} \frac{d\beta_n^*}{\sqrt{2\pi}}.$$

The normalizing condition (5.27b) for the vacuum functional Ω^0 requires that $\omega^0\{\zeta^*, \zeta, I\} = 1$ for $\zeta^* = \zeta = I = 0.$ Hence

$$N = \int \exp\left[-i \sum_{nm} \beta_n Q_{nm} \beta_m^* + iW_M\right] d(\beta)\, d(\Phi). \tag{6.21}$$

The calculation of integrals over a Fermi field, i.e., over β_n^* and β_n does not present now any special difficulties, as compared with integrals over a Bose field; the transition from (6.17) to (6.19) differs from the functional Fourier transformation from (5.5) to (5.25), by the fact that in (6.19) no operator Fourier transformation has been performed for the Fermi field*.

The operators χ, $\bar{\chi}$ with respect to the functional (6.19), will be of the form

$$\left.\begin{aligned} \chi(x) &= \sum_n \chi_n(x) \left(i \frac{\partial}{\partial \zeta_n^*}\right); \\ \bar{\chi}(x) &= \sum_n \bar{\chi}_n(x) \left(i \frac{\partial}{\partial \zeta_n}\right) \end{aligned}\right\} \tag{6.22}$$

The application of the operators χ and $\bar{\chi}$ to the functional (6.17), in accorddance with (6. 22), introduces into the integrand the additional factors β_n^* and β_m. It follows from the form of $\Omega\{\beta^*, \beta, \Phi'\}$ that, for $\zeta = \zeta^* = 0$ the integrals over β_n^* and β_m of the products of the coefficients β_n^* and β_m by $\Omega\{\beta^*, \beta, \Phi'\}$ will be different from zero only if these coefficients enter in parts $\beta_n \beta_n^*$. This corresponds to the fact that in the T-functions (5.22), each operator χ ought to enter only together with $\bar{\chi}$.

By using condition (6.12) (or (6.13)), we can now write the vacuum functional F_0 in the definition of the T-functions:

$$T(x\ldots|y\ldots|z\ldots) = [\chi(x)\ldots\bar{\chi}(y)\ldots\Phi(z)\ldots]\,\omega\{\zeta^*, \zeta, I\}, \tag{6.23}$$

where

$$\Phi(z) = i \frac{\delta}{\delta I(z)},$$

$\chi(x)$ and $\bar{\chi}(y)$ are determined by formulas (6.22), and after differentiation in (6.23), we have to put $\zeta = \zeta^* = I = 0$. Formulas (6.19)-(6.23) and condition (6.13) are basic for the calculation of the T-functions.

Let us consider the integration over a Fermi field. We calculate, as an example, the exact nucleon Green's function, which is, by (6. 23), equal to

---

* Detailed calculations of $\Omega\{\eta, \bar{\eta}, I\}$, by means of such an operator transformation, are given in reference /38/.

$$S'_F(x, y) = \chi(x)\chi(y)\omega\{\zeta^*, \zeta, 0\}|_{\zeta=\zeta^*=0}. \tag{6.24}$$

We introduce the notation

$$N(\Phi') = N\int \Omega\{\beta^*, \beta, \Phi'\}\, d(\beta); \quad \int N(\Phi')\, d(\Phi') = N. \tag{6.25}$$

Hence we can represent the expression for S'_F in the form

$$S'_F(x, y) = \frac{1}{N}\int S_F(x, y, \Phi')\, N(\Phi')\, d(\Phi'), \tag{6.26}$$

where, as we shall see, $S_F(x, y, \Phi')$ is the Green's function for the nucleon in the external meson-field Φ'. Indeed, from formulas (6.22) and (6.26) the expression for $S_F(x, y, \Phi')$

$$S_F(x, y, \Phi') = \sum_{l,m} \Psi_l(x)\overline{\Psi}_m(y) A_l B_m \int \beta_l^* \beta_m \exp\left[-i\sum_n \beta_n \beta_n\right] \times \\ \times \prod_n \frac{d\beta_n\, d\beta_n^*}{2\pi} = -i\sum_n \chi_n(x)\bar{\chi}_n(y) = -iD^{-1}(x, y, \Phi'), \tag{6.27}$$

where in integrating we put $Q_{nm} = \delta_{nm}$.

Thus the exact nucleon Green's function is obtained from the Green's function in the outer field $S_F(x, y, \Phi')$ by means of averaging over all the external fields with the weight functions $N(\Phi')$.

The quantity $N(\Phi')$ is also relatively easily transformed by means of the above discussed method. By (6.25)

$$N(\Phi') = \int \exp\left[-i\sum_{nm} Q_{nm}\beta_n\beta_m^* + iW_M\right] d(\beta). \tag{6.28}$$

By choosing in the normalizing condition (6.13) the Dirac operator $-i\delta^4(x-y)$ $D(y)$ without external field for $Z(x, y)$, we obtain

$$Q_{nm} = -i\int \bar{\chi}_n(x)\,[D(x) - g\gamma_5\Phi'(x)]\,\chi_m(x)\, d^4x = \\ = \delta_{nm} + ig\int \bar{\chi}_n(x)\,\gamma_5\,\Phi'(x)\,\chi_m(x)\, d^4x \equiv \delta_{nm} + igq_{nm} B_n A_m. \tag{6.29}$$

If B_n and A_m were not anticommuting operators but numbers, i.e., if the operators χ and $\bar{\chi}$ referred to the Bose field, we would obtain that

$$N_B^0(\Phi') = e^{iW_M} N(\Phi) = \int \exp\left[-i\sum_{nm} \beta_n\beta_m^*(\delta_{nm} + igq_{nm})\right] d(\beta) = \\ = \int e^{-i\sum_n \lambda_n\lambda_n^*} \frac{\partial(\beta, \beta^*)}{\partial(\lambda, \lambda^*)}\, d(\lambda). \tag{6.30}$$

A transformation of the variables $\beta, \ \beta^*$ to the variables $\lambda, \ \lambda^*$ is a transformation which diagonalizes the quadratic form $\Sigma \beta_n \beta_n (\delta_{nm} + igq_{nm})$. The Jacobian of this transformation is equal to the inverse determinant, formed by the coefficients of this quadratic form

$$\frac{\partial (\beta, \beta^*)}{\partial (\lambda, \lambda^*)} = || \delta_{nm} + igq_{nm} ||^{-1} \tag{6.31}$$

and does not depend on λ, λ^*; hence

$$N_B^0 = || \delta_{nm} + igq_{nm} ||^{-1}.$$

In the case when χ and $\bar{\chi}$ refer to a Fermi field, we have to diagonalize the form $\Sigma \beta_n \beta_m^* (\delta_{nm} + igq_{nm} B_n A_m)$ and we obtain by analogy that

$$N^0 (\Phi') = || \delta_{nm} + igq_{nm} B_n A_m ||^{-1}.$$

In the case of infinite determinants of the Fredholm type, the inverse value of the determinant $|| \delta_{nm} + gG_{nm} ||$ is equal to the permanent of the form $(\delta_{nm} - gG_{nm})$, i.e., to a quantity which is calculated according to the same rules as the determinant, but without change of signs, i.e., only with positive signs

$$N^0 (\Phi') = \mathrm{Perm}\, (\delta_{nm} - igq_{nm} B_n A_m).$$

If in the expansion for the permanent we commute the operators B_n and A_n, so as to use formula $B_n A_n \to 1$, this will lead to a change of sign which in fact compensates the change of sign in the transition from the permanent of the determinant /45/

$$N^0 (\Phi') = || \delta_{nm} + igq_{nm} ||. \tag{6.32}$$

The same result can be obtained directly from formula (6.28), after the substitution of Q_{nm} from (6.29), the expansion of $\left[-g \sum_{nm} q_{nm} \beta_n \beta_m^* B_n A_m \right]$ in a series, and the term-by-term integration with the use of the equality $B_n A_n = 1$.

The meaning of (6.32) can be easily elucidated if we reduce the matrix q_{nm} to the diagonal form, where

$$q_{nm} = \delta_{nm} q_n;$$

$$N^0 = \prod_n (1 + igq_n),$$

or

$$\ln N^0 = \mathrm{Sp} \ln (1 + ig\hat{q}).$$

Since

$$q_{nm} = \int \overline{\psi}_n(x) \gamma_5 \Phi'(x) \psi_m(x) d^4x$$

and, by (6.13) $\Sigma \Psi_n(x) \overline{\Psi}_n(y) = -iS_F(x-y)$, then $\ln N^0(\Phi')$ can also be represented in the form

$$\ln N^0(\Phi') = \mathrm{Sp} \ln (1 - gS_F \gamma_5 \Phi), \tag{6.33}$$

if we consider $S_F(x, y)$ is matrix elements of the operator S_F. A further transformation of (6.33) leads to (see, e. g., /3/) the expression

$$N^0(\Phi') \exp \left\{ -g \, \mathrm{Sp} \, \gamma_5 \int_0^1 d\lambda \int S_F(x, x, \lambda\Phi') \Phi'(x) d^4x \right\}. \tag{6.34}$$

Formula (6.26) for S'_F, as well as formula (6.34), were deduced by different methods /43, 44, 45/. The derivation given illustrates the method of direct integration over a Fermi field. Usually, integration over a Fermi field is carried out first, whereas approximation methods of functional integration are associated with the study of integrals over a Bose field /45/. The possibility of reducing the integrals over anticommuting functions to integrals over both fields on the same level may possibly be convenient for the investigation of the difficulties of the present-day theory.

BIBLIOGRAPHY

BIBLIOGRAPHY

1. Fock, V. A., Zeit. f. Phys. 49, 339, 1928.

2. Fock, V. A., Phys. Zs. d. Sow. Union 6, 425, 1934; Vestnik LGU, (News of the Leningrad State University), 3, 108, 1937.

3. Berestetskii, V. B., Galanin, A. D., referativnyi sbornik "Problemy sovremennoi fiziki" (Coll. "Problems of Modern Physics") 3, 1955.

4. Silin, V. P., Feinberg, E. Ya., UFN 56, No 4, 1955.

5. Smirnov, A. A., ZhETF 5, 687, 1935.

6. Vlasov, A. G., ZhETF 10, 1151, 1940.

7. Fedorov, F. I., Uchennye zapiski LGU seriya fiz. nauk, (Scientific Notes of the Leningrad State University, physical science series), No 146, 8, 1942.

8. Dirac, P. A. M., Proc. Irish. Roy. Soc.

9. Schwinger, J., Proc. Nat. Acad. Sci. 37, 452, 455, 1951; see also: "Problems of Modern Physics", No 3, 1955.

10. Feynman, Rev. Mod. Phys. 20, 367, 1948.

11. Friedrichs, Mathematical aspects of the Quantum Theory of Fields. Inter. Pub., New York, 1953.

12. Symanzik, K., Zs. f. Naturforschung 9a, No 10, 1954.

13. Fock, V. A., Zs. f. Phys. 75, 622, 1932.

14. Tamm, . E., Journ. of Phys. 9, 445, 1945.

15. Dankoff, S. M., Phys. Rev. 78, 382, 1950.

16. Cini, M., Nuovo Cimento 10, 526, 614, 1953.

17. Levy, M., Phys. Rev. 88, 72, 725, 1952.

18. Klein, A., Phys. Rev. 90, 1, 101, 1953.

19. Lehmann, H. Zeits. f. Nat. 8a. 579, 1953.

20. Pirenne, I.. Physica XV, No 11 12, 1, 023, 1949.

21. Dyson, F., Ross, M., Salpeter, E.E., Schweber, S.S., Sundaresan, M.K.', Visscher, W.M., Bethe, H.A., Phys. Rev. 95, 1644, 1954.

22. Dyson, F., Phys. Rev. 91, 1543, 1953.

23. Dalitz, R.H., Dyson, F.I., Phys. Rev. 99, 301, 1955.

24. Wightman, A.S., Schweber, S.S., Phys. Rev. 98, 812, 1955.

25. Novozhilov, Yu.V., ZhETF 22, No 3, 1952; DAN 83, No 2

27. Schwinger, J., Phys. Rev. 91, 713, 1953; 91, 728, 1953; 92, 1283, 1953; 93, 615, 1954; 94, 1362, 1954.

28. Matthews, P.T., Salam, A., Proc. Roy. Soc. A221, 128, 1954.

29. Feynman, R.P., Phys. Rev. 80, 440, 1950; 84, 108, 1951.

30. Landau, L.D., Pomeranchuk, I.Ya., DAN SSSR 102, 489, 1955; Bogolyubov, N.N., Shirkov, D.V., DAN 115, No 4, 1955.

31. Heisenberg, W., Nachr. d. Gott. Adak. d. Wissensch., No 8, 1953.

32. Lehmann, H., Symanzik, K., Zimmerman, W., Nuovo Cimento 1, No 1, 1955.

33. Katayama, Y., Tokuoka, Z., Yamazaki, K., Nuovo Cimento 2, 728, 1955.

34. Valatin, J., Proc. Roy. Soc. A229, No 1177, 1955.

35. Novozhilov, Yu.V., DAN 104, 47, 1955; ZhETF 31, 493, 1956.

36. Matthews, P.T., Salam, A., Nuovo Cimento 2, 120, 1955.

37. Coester, F., Phys. Rev. 95, 1318, 1954.

38. Gol'fand, Yu.A., ZhETF 28, 140, 1955.

39. Preese, E., Zeits. f. Naturf. 8a, 776, 1953; Nuovo Cimento 11, 312, 1954.

40. Nishijima. K., Theor. Phys. 10, 549, 1953; 12, No 3, 1954.

41. Akhiezer, A., Berestetskii, V. B., Kvantotaya elektrodinamika, Gostekhizdat, 1954 (Quantum Electrodynamics).

42. Gel'fand, I. N., Minlos, R. A., DAN SSSR 97, 209, 1954.

43. Fradkin, E. S., DAN SSSR 98, 47, 1954; ZhETF 29, 121, 1955.

44. Bogolynbov, DAN SSSR 99, 225, 1954.

45. Edwards, S. F., Proc. Roy. Soc. 232A, 371, 377, 1955.

46. Klepikov, N. P., DAN SSSR 98, No 6, 1954.

47. Joffe, B. L., DAN SSSR 95, 761, 1954; Galanin, A. D., Joffe, B. L., Pomeranchuk, I. Ya., DAN SSSR 98, 361, 1954.

48. Lehmann, H., Nuovo Cimento 11, 342, 1954

49. Zimmermann, W., Suppl. Nuovo Cimento 11, 43, 1954.

50. Symanzik, K., Zeits. f. Nat. 9a, 809, 1954.

51. Freese, E., Nuovo Cimento 2, 50, 1955.

52. Watanabe, I., Prog. Theor. Phys. 10. 371, 1953.

53. Edwards, S. F., Phil. Mag. 47, 758, 1954.